Fashionable Clothing from the Sears Catalogs

Mid 1960s

Schiffer Publishing Ltd

4880 Lower Valley Road, Atglen, PA 19310

Joy Shih

Sears

*Everything for
Spring through Summer
1966*

This dress is from
our all-new
Junior Department
PAGES 18-51

Index
begins on page 635

SEARS, ROEBUCK
AND CO.
GREENSBORO, N. C. 27402

Satisfaction guaranteed
or your money back
YOU CAN COUNT ON US

D0860842

To my sister Gloria,
thanks for catalog "dreaming" with me.
Remember Dad got mad about the clothes under our beds?

Thanks to
Desire Smith, for your expertise in
the area of collectible clothing.
Tammy Ward, for your tireless assistance.
Thanks Blair, you're the best!

Copyright © 1997 by Schiffer Publishing
Library of Congress Catalog Card Number: 97-80088

ISBN: 0-7643-0340-6
Printed in Hong Kong

Book layout by: Blair Loughrey

Published by Schiffer Publishing Ltd.
4880 Lower Valley Road
Atglen, PA 19310
Phone: (610) 593-1777; Fax: (610) 593-2002
E-mail: Schifferbk@aol.com
Please write for a free catalog.
This book may be purchased from the publisher.
Please include $3.95 for shipping.
Please try your bookstore first.

We are interested in hearing from authors
with book ideas on related subjects.

Title page photo: Cover photo featuring teen model Cheryl Tiegs, from the Spring and Summer 1966, Greensboro, North Carolina Edition 232, ©Sears, Roebuck, and Co.

Contents

Introduction   5

Women's Fashions   11

Formal Wear   11
Day and Career Wear   22
Casual Dresses   55
Casual Separates   58
Sportswear   69
Beachwear   74
Maternity Wear   78
Loungewear and Sleepwear   82
Fashion Accessories   87
Suits   96
Outerwear   103
Half Sizes   112

Teen Girls' Fashions   117

Dresses   117
Casual Separates   121
Sportswear and Beachwear   123
Outerwear   128

Men's Fashions   130

Suits and Sports Jackets   130
Sportswear   133
Sleepwear   137
Outerwear   138

Children's Fashions   139

Younger Children   139
Girls' Wear   141
    Dresses   141
    Separates   148
    Sportswear   151
    Outerwear   156
Boys' Wear   157
    Dress Wear   157
    Sportswear   158
    Outerwear   160

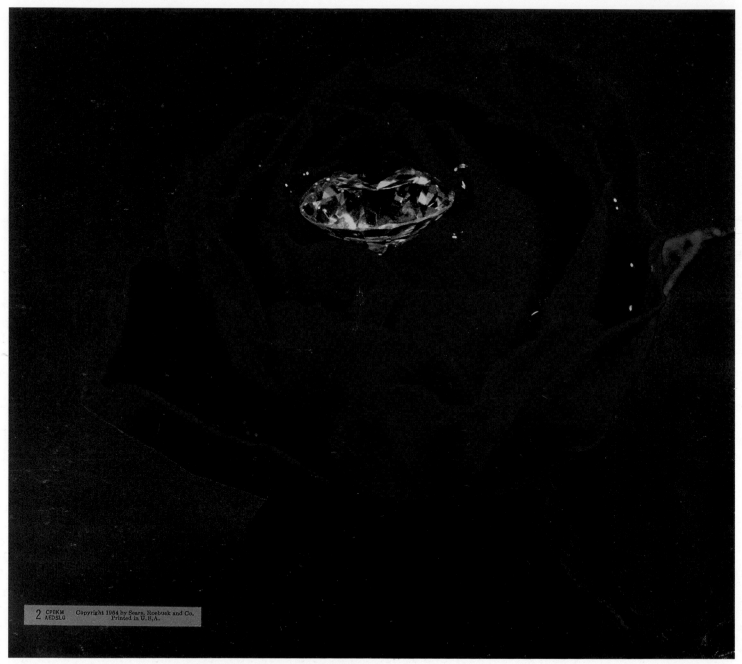

"Fashion is forever at Sears." A solitaire engagement ring, brilliant cut
premium quality diamond, one-piece platinum mounting. 1-carat, $895.00.
2-carat, $2500.00. 3-carat, $6500.00. [NPA] Spring/Summer 1964.

Introduction

Looking through the Sears Catalog is like walking through a door into American's past. The images within not only give clues as to what the average person wore during that era, but tell us a great deal about American culture.

In the 1960s, the Sears Catalog, as it had for the previous 80 years, remained the average person's fashion magazine. Before the advent of suburban shopping malls, the only places to purchase fashionable clothing were in larger city department stores. Many people living in rural areas did not have ready access to these stores and depended on the Sears Catalog to keep their families dressed in the current fashion styles. Sears brought the latest fashions from New York, California, or London to Main Street, USA.

In this book, you are invited to look at the Sears catalogs through different eyes. Informative descriptions of each item give an idea into the kind of clothing styles popular from 1964-1966, color trends for each season, the "newest" fabrics, the cost of the item in the 1960s, and if available, the current market values of the items as collectible clothing.

About the Sears Catalog

Prior to and up until the mid 1960s, Sears department stores sold fashionable clothing only through their catalogs. Orders were placed by mail, telephone, telegraph or placed in person at a Sears Catalog Sales Office, Retail Store Catalog Sales Department or in one of Sears affiliated stores. Shopping from the Sears Catalog was easy, convenient, and safe. Sears offered a satisfaction guaranteed or your money back policy, as it had since the 1880s. "It's our promise of honest dealing. It's your assurance of good merchandise and good value."

By the mid 1960s, the Sears Catalog showed most fashions in color. Customers were assured that the actual colors of the merchandise were reproduced as faithfully as was possible with the latest "modern high-speed printing equipment." The special care to reproduce exact and accurate colors was important to back up the company's unconditional return guarantee.

Besides the fashion catalog, specialty catalogs were available including ones for Business Equipment and Supply, Custom Tailored Clothing, Suburban Farm Supplies and Fencing, and a Decorating Book. In addition to domestic catalogs, Sears also offered catalogs that could be used in over 100 countries around the world. These catalogs contained the same merchandise, at the same prices as the regular catalogs, but included special ordering information to make overseas shopping easier. As in the past, Sears continued to take orders in any language at any of their eleven Catalog Order Plants around the United States. During the mid 1960s, catalog ordering information included instructions for U.S. military personnel ordering from overseas, as well as employees of American companies in foreign residence, missionaries, and students. Sears shipped most merchandise overseas but could not export "live animals or poultry, guns or ammunitions, drugs or nursery stock." We knew that Sears sold a lot of things, but live animals and poultry?

We learned from the catalogs of 1964-1966 that Sears received payment by post office money orders, express money orders, bank draft, or check. A company-issued credit card that required no money down and up to five years to pay at 1-1/2% interest was an option.

Merchandise was delivered to the customer by parcel post, and larger items by truck. All bird and animal pets, as well as bulky, unmailable items were shipped by railway express on fast passenger trains. Parcel post rates in 1966 were 4 cents for the first 2 ounces and 2 cents for each additional ounce. A five pound item shipped by railway express to a destination 100 miles away cost about $3.30. Sears credit card customers had the added advantage of having the shipping charges figured out for them.

For convenience and the assurance of wearing the latest fashionable clothing, the Sears Catalog remained the best way to shop.

Sears Catalog as Fashion History

Fashion in the mid 1960s presented a dramatic difference from the 1950s. For the first time, clothing was less formal. The number of catalog pages devoted to sportswear showed that people were more active. Even the models were posed in active positions, with legs outstretched and bodies twisted simulating movement. Convenience and comfort became selling points, spawning products such as the combination wig-hat that seems ridiculous to us today.

Women's clothing, perhaps to accommodate more active lifestyles, became less fitted than in years past. Waistlines rose to just below the bustline in the popular empire-waist dresses. A-line and slim pleated skirts replaced the "unpressed pleats" look in full swing skirts. Hemlines fell just at the knees, and slightly above for sportier designs.

Considered a good idea at the time, this combination wig and hat was the answer to a bad hair day. "Warm enough to wear outside, comfortable enough to be kept on indoors." Made of fluffy modacrylic pile on acrylic backing, these wig-hats were easily styled and came in exciting shades of light auburn red, platinum blonde, honey blonde, and medium brown (not shown). Sent styled as in the center illustration, women were encouraged to "have fun creating your own" hairdo. $4.97. [$12-18] Fall/Winter 1965.

The British Invasion of music also popularized the slim hip-hugger pants look as well as the stylish Chelsea collar and "bobby-styled" hats. Capes were popular coverups. Triangle kerchiefs or scarves were must-have accessories. Jewelry was simple and minimal, and pins were worn high almost near the shoulder.

Just as western-style clothing was going out of style for men and children, women's clothing started to feature the "frontier" look with double top stitching detail and western-style yokes. The bell-bottom pants started to come into style at this time. Although bell bottoms would continue to be in style, in its various forms, up and until the late 1970s, mid 1960s bell bottom pants were fitted through the legs and flared below the calves.

Fabrics prevalent in women's fashions included the flocked dotted Swiss, loopy textured knits, gingham checks with lace, denim, calico prints, and the ever-present madras plaid. Genuine madras from India, colored with vegetable dyes, tend to bleed into soft hues. American textile manufacturers took advantage of the popular madras craze by making madras-look plaids that did not bleed. Madras was seen in clothing worn by men, women, and children.

Men's slacks were more fitted and hugged close to the legs. The British influence provided slimmer, almost tight pants and slim pointy shoes. The California surfer look brought bright tropical "jam" prints, henley collar sport shirts, and madras windbreakers. The Jac-shirt, combining a light jacket and a shirt, was still fashionable but with a shorter, more fitted look than its late 1950s counterpart. Sport jackets were "dressed down" with casual seersucker and plaid design fabrics, and featured more casual styling when worn with an open collared shirt and an ascot.

A madras plaid shirt was a must for the fashionable teen of the mid 60s. Spring/Summer 1966.

Slim tapered shoes for men include the high-rise oxford, the alligator print glossy black leather, and the "newsworthy high British heel" slip-on boot. $9.77 pair. [$45-65] Spring/Summer 1965.

Sears Catalog as Cultural History

For the new generation of "baby boomers" now coming of age in the mid 1960s, the catalogs offered much more than clothing. By being able to emulate styles worn from coast to coast, any teen across the country could be a part of the growing youth culture. From madras plaids to bright empire shifts, the American teen was suddenly the focus for merchandising. Catalogs in the mid 1960s began to show teens on the cover. For the first time, the catalogs dedicated a large part of each edition to Junior fashions for teen girls. By 1965, the first portion of each catalog started out with these fashions, relegating older women's fashions to a spot after Junior fashions.

Women in the mid 1960s were beginning to get out of the house. The catalogs showed much less clothing as stay-at-home loungewear, and much more as day and career dresses. The growing youth culture also helped to make clothing less formal, with styles featuring bolder, cleaner lines. For the first time, denim came out of the workplace as rugged "factory wear" and into the wardrobe of everyday people. The use of denim in clothing was one way to "dress down," and with the addition of ruffles and lace, lend a youthful look to fashion.

An interesting addition during this period was a section of each catalog specifically targeted for "Shorter Women Who Wear Half Sizes." These fashions were sized with slightly higher waistlines, eased bodices and skirts, roomier sleeves but "never wide shouldered." Apparently, clothing manufacturers were beginning take notice and accommodate women of all sizes.

The Sears Catalog during the mid 1960s, as compared to previous years, was beginning to show a more youthful culture and a more leisurely lifestyle. Casual fashions and sportswear grew in numbers. Models posed in outdoor recreational settings, at the beach or at the ranch. The power of advertising pictured society as youthful, healthy, and active.

Sears Catalog as Nostalgia

A first glance at the Sears Catalogs from 1964-1966 shows some familiar faces. Some of the young teen models pictured on these pages later became major television and media celebrities. Perhaps the increasing youth population at the time help to propelled these persons to stardom. More youth meant more teen magazines, music, and television coverage. After all, every teenage girl wanted to dress like model Cheryl Tiegs.

My sister and I use to leaf through the pages of the catalogs, adding up all the clothing we liked, and ending up with totals in the thousands. As if we could afford it, we would eventually eliminate items to an affordable amount in the "hundreds." It was, after all, not just a game, but a dream to be the best dressed girl in town.

Certainly for "baby boomers" the fashions on these pages will be familiar. Ruffled lace tops, gingham check shifts, madras jackets, and hip-hugger bellbottoms all played a part in our growing up years. Similarly, for those who were not around during the mid 1960s, the styles that have survived the test of time will be here too. In addition, how many celebrities do you recognize on these pages?

Sears Catalog as a Collectibles Guide

All items pictured in this book are from the mid 1960s and are of interest to collectors and dealers of *wearable* vintage fashions. Highest prices are paid for items that are rare, or show originality of design and construction. The traditional classic styles of the 1960s, such as the shirtwaist dress, have little or no market value. 1960s vintage, however, such as bell bottoms, hip-huggers, and stretch knits in psychedelic colors are highly sought after by collectors. Junior fashions are considered more collectible than styles for older women.

Fabric also affects price. Oriental beadwork, pure linen, silks, and cashmere are all marketable. Style is most important to collectors. Black cocktail dresses, especially the low-cut styles, are always in demand. The beautiful silk prints are beginning to be of interest to collectors.

Items that are of special interest to designers are swimwear and sleepwear, mainly for design inspiration. Specialty items such as baseball uniforms, "space" suits, and Roy Rogers outfits bring high prices because they cross over from vintage fashions into the collectibles market.

Regional differences, trends in the market, and selling venue dramatically affect prices of vintage clothing and accessories. Highest prices are always paid for items in excellent condition. Similar items in poor condition may have little or no value. Current market values in this book are listed inside brackets after 1960s prices, and are for items in excellent condition.

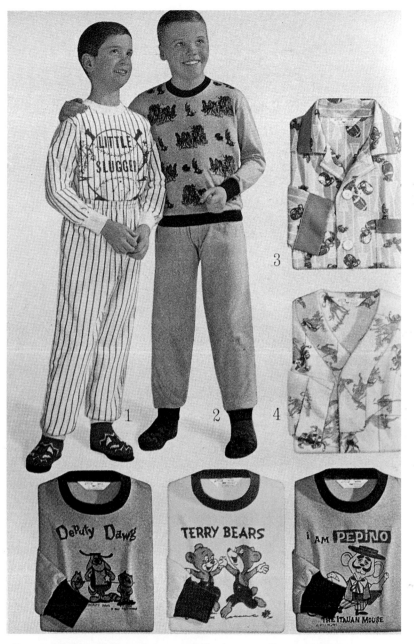

Cotton pajamas in flannelette or combed cotton. Shown in "little slugger", stagecoach, football, rodeo, and cartoon styles. Cartoons feature Deputy Dawg, Terry Bears, and "Pepino" Mouse. $1.99-$2.54. [$20-25] Fall/Winter 1964.

Women's Fashions

Formal Wear

Late-day dress of rayon knit, back-tied drawstring. "Unpacks without wrinkles." Black. $26.00. [$55-60] Spring/Summer 1964.

Slim dance dress in lightweight Zefkrome® acrylic double-knit. Lined bodice, deep back slit. Gray and white. $17.90. [$35-40] Spring/Summer 1964.

Draped long dress in matte jersey of acetate and rayon, fully lined. Black or white. $30.00. [$65-70] Spring/Summer 1964.

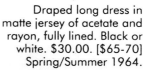

Romantic lace dress in two lengths. White acetate lace over acetate taffeta, scalloped at the hemline for a border effect. Skirt fall in full folds in back, with a taffeta bow with long streamers. Attached nylon net and cotton petticoat. Floor length in all white, $23.50. [$45-50] Street-length in white over aqua, $19.50. [$35-40] Spring/Summer 1964.

Black Sheath with Silver-colored Bead Trim. Acetate and rayon crepe, fully lined. Also in pink. Street length, $23.00. [$55-60] Floor length, $25.00. [$35-40] **Lace and Net Shift.** Nylon/acetate overlaid with rayon ribbon. White. Also in black. $20.00. [$45-50] **Two-piece Lace Dress.** Cotton/acetate lace over acetate taffeta. Sheath zips in back, overblouse buttons in back. Blue. Also in white. Street length, $23.00. [$35-40] Floor length, $25.00. [$45-50] Spring/Summer 1964.

Two-piece Peplum Sheath Dress. Rayon chiffon over acetate taffeta sheath, tiny straps, cotton lined bodice. Sleeveless overblouse has back opening. Red. $17.84. [$35-40] **Two-piece At-home Entertaining Dress.** Sleeveless black top of rayon matte jersey, lined in acetate crepe. Black and white floral print floor length skirt in linen-look rayon, lined in acetate taffeta. $19.84. [$25-30] **Crystal Pleated Dress.** Rayon chiffon over acetate, midriff in black cotton/silk/rayon lace all around. Black. $19.84. [$65-70] Spring/Summer 1964.

Lace Ruffle Jacketed Dress. Black acetate and rayon crepe dress, fully lined. Lace ruffle jacket has pink camellia ornament. $18.00. [$55-60]
Lace Flounce Dress. Black acetate and nylon lace, fully acetate lined, attached nylon net petticoat. $22.00. [$55-60] Fall/Winter 1964.

After-five fashions. **Cardigan Suit.** Winter-white wool, mohair, and nylon accented with gold-color trim. Includes off-white acetate/rayon crepe sleeveless blouse. $30.00. [$30-35] **Hand-loomed Knit Wool Shell.** Hand-sewn iridescent beads and sequins. Silk-lined. Winter white. $29.90. [$55-60] **Long Skirt.** Loopy wool, mohair and other fibers. Fully lined. Winter white. $16.90. [$20-25] Fall/Winter 1964.

Junior dresses in a party mood. **Lace-top Dress.** Dacron® polyester batiste with acetate and nylon lace. Light yellow. $13.84. [$55-65] **Triple Tier-skirted Dress.** Polyester and cotton broadcloth dress with cotton lace and blue velvet trim. White and blue. $14.84. [$55-65] **Bolero Bouffant Dress.** Acetate and nylon lace bodice and button back bolero, sheer nylon skirt. Completely lined in acetate taffeta with an attached nylon net petticoat. Light rose pink. $16.84. [$55-65] Spring/Summer 1964.

13

Beautiful semi-formal silk dresses. **Bow-tied Shift.** Black and camel tan surah pattern with billowy raglan sleeves, completely lined. $25.75. **Poet's Collar Shirtdress.** Bright green and blue floral print broadcloth, bodice and skirt lined. $24.75. Fall/Winter 1964.

Emerald Green Party Dress. Belled skirt is Pellon®-lined, bodice is acetate lined. Rayon velvet. $23.50. [$65-70] **Dark Blue Ball Gown.** Sparkling rhinestones and beads circle waist. Rayon, fully lined. Floor length, $29.50. [$95-110] Street length, $26.50. **Ruby Red Theatre Suit.** Jacket and skirt of rayon velvet, fully lined. Acetate satin pullover pale pink blouse has side zipper. $28.50. [$85-95] Fall/Winter 1964.

Brocaded palace silks imported from Hong Kong. **Frog-trimmed Two-piece Dress.** Hand-bound buttonholes and covered buttons at back closing. Completely lined. Peacock blue. $26.75. [$50-55] **Jacket Dress.** Jacket with frog-closed pockets, back closing with covered buttons and hand-bound buttonholes. Fully lined. Emerald green. $29.75. [$55-60] Fall/Winter 1964.

Glittering touches add glamour to these elegant semi-formal dresses. **Two-piece Textured Wool Dress.** Rhinestone sunbursts detail overblouse. Fully lined. White. $19.84. [$55-65] **Clutch Bag.** Alligator-skin plastic, rayon lined. Alabaster - oyster white. $3.77. [$30-35] **Misty Rayon Chiffon Dress and Jacket.** Jacket of Mylar® metallic, acetate, and nylon. Bow-trimmed front hooks, fully lined. Light beige. $23.84. [$60-65] Fall/Winter 1964.

Velvet and Brocade Gown. Sleeveless black rayon velvet bodice, brocaded white cotton and acetate skirt. Rhinestone trim on pink acetate satin belt. Fully lined. Floor length, $19.84. [$40-45] Street length, $17.84. [$35-40] **Lace Sleeved Sheath.** Black acetate and rayon crepe, white nylon and acetate lace sleeves overlaid with rayon ribbon. Dress fully lined. $13.84. [$75-80] **Bright Rose Sheath.** Rayon and silk shantung, lined except sleeves. Pin not included. $12.84. [$35-40] **Cotton Lace Dress with Overshift.** Acetate and rayon crepe sheath with spaghetti straps, fully lined. Overshift is cotton lace. Each with long back zipper. Pale pink. $16.84. [$55-60] Fall/Winter 1964.

Lace Suit Dress. Acetate and nylon lace with scalloped edging on jacket, cuffs, and skirt. Fully lined in acetate taffeta. Beige. $12.84. [$50-55] **Empire Waist Slim Sheath.** Twill-back cotton velveteen. Turquoise blue. $11.54. [$20-25] **Brocade Dress.** Cotton and acetate brocade with scallop trim and bows. Light blue. $10.84. [$20-25] **Two-piece Brocade Dress.** Detachable rayon chiffon scarf. Overblouse fully lined. Green and Blue. $12.84. [$15-20] Fall/Winter 1964.

15

Junior semi-formal party dresses. **Lace and Chiffon Dress.** Acetate and nylon lace bodice and chiffon skirt, lined. Sleeve piping and belt of rayon satin. Light blue. $14.84. [$60-65] **Schiffli-embroidered Dress.** Acetate and rayon faille bodice scoops to low V in back. Rayon embroidery circles the nylon organdy skirt. Waistband with bow in rayon velvet. Dress completely lined in acetate taffeta. Black and white. $13.84. [$55-60] **White Eyelet Dress.** Schiffli-embroidered white eyelet cotton dress with cotton sateen waistband and posy trim. White with barest pink. $13.84. [$55-60] Spring/Summer 1965.

White lace evening or wedding gown with a beaded and scroll-trimmed midriff band. Skirt of billowy sheer nylon with attached nylon net petticoat. Lined in acetate taffeta. Also available sized for shorter women who wear half sizes, with higher waistlines, fuller bodices and sleeves. $23.50. [$50-55] Fall/Winter 1964.

Tiers of Lace Gown. White tiers of acetate and nylon lace on rayon and silk organza. Dress lined in acetate taffeta. Mint green. Floor length, $21.84. [$55-75] Street length, $18.84. [$45-65] **Embossed Two-Piece Gown.** Embossed acetate with the look of silk. Bolero jacket with back closing. Bodice and jacket lined in acetate taffeta, skirt lined in Pellon® with attached petticoat. Floor length, $19.84. [$50-55] Street length, $17.84. [$40-45] Spring/Summer 1965.

16

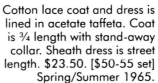

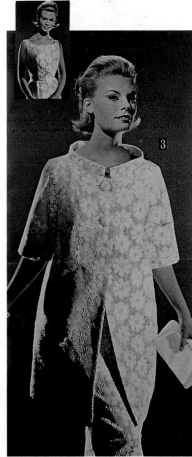

Cotton lace coat and dress is lined in acetate taffeta. Coat is ¾ length with stand-away collar. Sheath dress is street length. $23.50. [$50-55 set] Spring/Summer 1965.

"Dancing Crepe" black dresses, worn a little shorter than average dresses for ease of movement. Made of acetate and rayon, these styles are completely lined. **V-neck with Bolero Effect Bodice.** Price included a 5-cent Federal Excise Tax on rhinestone pin. $15.84. [$50-55] **Scooped Neckline.** Double ruffle flounce at hem. $15.84. [$50-55] **Petal-trimmed Neckline.** A-shape dress. $17.84. [$60-65] **Textured Pantyhose.** "No garters show as skirts swing, just dance-pretty legs." Shown in Smoketone. $2.87, pair. [NPA] Spring/Summer 1965.

Empire Ruffle and Lace Dress. Scalloped lace of rayon, cotton, and nylon, lined in acetate taffeta. Rayon velvet bow. White. $20.00. [$85-90] **Big Patterned Lace Ball Gown.** Midriff and lining in acetate taffeta. Skirt is Pellon®-lined to retain shape. White on pink. $22.50. [$40-45] Spring/Summer 1965.

17

Brocade Bell Dress. Acetate brocade with bell-shaped skirt. Neckline dips to "V" in back. Lined. Pale sky blue. $19.84. [$40-45] **Glittery Gown.** Bodice of bright gold-color rayon, Mylar® metallic and cotton. Slender skirt is cream-white acetate and rayon crepe. Gold acetate satin cummerbund has bow and streamers in back. Fully lined. Walking slit in back. $20.84. [$40-45] Fall/Winter 1965.

Rich rayon and acetate crepe dress with cape-effect drapery in back. Bodice front features a slightly raised waist. Fully lined. Rose pink. Street length, $17.84. [$40-45] Floor length, $19.84. [$50-55] Spring/Summer 1965.

18

"Lady Fauntleroy" Suit Dress. Ruffle blouse, quilted jacket, and skirt of acetate crepe. Jacket lined, skirt seat-lined. Light yellow gold. $14.54. [$45-50] **Brocade Belle Dress.** Cotton and rayon, lined. Gold. $10.94. [$40-45] Fall/Winter 1965.

Scallop Lace Coat and Dress.
Acetate and nylon lace sleeveless
coat with a mandarin collar.
Acetate and rayon crepe dress
with acetate satin trim, fully lined.
Light sky blue. $16.84. [$45-50]
Fall/Winter 1965.

Chartreuse Chic Dress.
Acetate and rayon matelassé
crepe, rayon chiffon scarf.
Lined. Chartreuse green.
$13.84. [$40-45] Fall/Winter
1965.

**Romantic Two-tone Chiffon
Dress.** Rayon chiffon, lined in
acetate taffeta. Neckline also V-
shaped in back. Fuchsia with pale
pink. $17.84. [$50-55] Fall/Winter
1965.

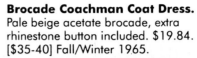

Brocade Coachman Coat Dress.
Pale beige acetate brocade, extra
rhinestone button included. $19.84.
[$35-40] Fall/Winter 1965.

Barest Illusion Lace Dress.
"Attract him in this sheath that
cleverly combines black cotton,
acetate, and nylon lace with black
acetate and rayon crepe." Bodice
lined in deep beige acetate
taffeta, skirt also lined. $19.84.
[$50-55] Fall/Winter 1965.

Sweet you

1

Junior size spring dance dresses. **Schiffli-embroidered Checks.** White linen-look rayon top, woven polyester and cotton check skirt, topped with an embroidered rayon and silk organza overskirt. Rayon grosgrain ribbon trim at waist. Copen blue and white. Long length, $19.90. [$40-45] Street length, $17.90. [$30-35] **Floral Applique Empire Dress.** Rayon chiffon dress and pop-over jacket, lined in acetate taffeta. Acetate satin ribbon band and bow. Light pink and white. Long length, $22.90. [$40-45] Street length, $19.90. [$30-35] Spring/Summer 1966.

JUNIOR ROUTE 1966

Junior dresses in the party mood. **Lace and Chiffon.** Shift topped with illusion cotton lace, acetate taffeta lined. Pink. $18.90. [$40-45] **Pleated Dance Dress.** Nylon organza pleats on an acetate taffeta dress. Acetate satin bow and waistband. Bright turquoise blue. $16.90. [$25-30] **Big Flower Pattern Lace Dress.** White cotton lace over ivory acetate taffeta. Bow and buttons in matching taffeta. $17.90. [$40-45] Spring/Summer 1966.

Scallop Lace Gown. Acetate and nylon lace on acetate taffeta, lined. Neckline dips to a "U" in back. Pink with white. Floor length, $23.50. [$30-35] Street length, $21.50. [$35-40] Spring/Summer 1966.

Empired-styled Gown. Acetate and rayon crepe dress, fully lined in acetate. Attached sash, back bow, and streamers of shiny acetate satin. Pink. Floor length, $19.50. [$25-30] Street length, $17.50. [$20-25] Spring/Summer 1966.

20

Cowl-collared Dress. Fine rayon and acetate crepe, fully lined with rayon. Bright pink. $16.90. [$20-25] Spring/Summer 1966.

Two-piece Floral Lace Dress. Dress of acetate and rayon crepe has "spaghetti" straps, empire bodice, flared skirt, fully acetate-lined. Cotton lace overblouse is trimmed with dress fabric, buttons in back. $15.90. [$20-25] Spring/Summer 1966.

Skimmer. Cotton lace with silk shantung collar, cuffs and bow. Lined in acetate except for sleeves. Pink. $17.90. [$30-35]
Two-piece Ruffle Dress. Acetate, cotton, and nylon lace, lined in acetate except for sleeves. Pale beige. $15.90. [$40-45]
Ensemble. Coat and dress of cotton and nylon lace. Dress fully lined in acetate. Light blue. $23.50. [$30-35] Spring/Summer 1966.

Ruffled Party Dress. Misty-sheer rayon georgette, lined with acetate taffeta. Attached belt and bow of rayon satin. Light pink. $15.90. [$30-35] **Embroidered Blouson Dress.** Rayon chiffon two-piece, covered with soutache scroll trim of rayon and cotton. Fully lined in acetate taffeta. Fuchsia rose. $25.00. [$20-25] Spring/Summer 1966.

Pleated sleeveless dress in pale crepe of Arnel® triacetate and rayon, lined bodice, contoured waistband dips in back. Sear's *Claude Riviere Collection made in the USA*. White. $19.50. [$20-25] Spring/Summer 1964.

Women's Fashions
Day and Career Wear

Left:
Two-piece dress designed by Parisian Claude Riviere in a brilliant print of rose-pink, yellow and beige on linen-look rayon. A-line skirt has two back zippers for a smoother fit. Overblouse zips in back. Sear's *Claude Riviere Collection made in the USA*. $22.50. [$20-25] Spring/Summer 1964.

Soft-flowering print sheath, fully lined filmy silk chiffon with gently shirred neckline, bow trimmed belt, and separate stole to wear cape-like or draped cowl-fashioned. Pink/orange. $23.50. [$35-40] Spring/Summer 1964.

Pebble print silk jersey shift with a "new" high, zipper-closed convertible neckline. Lime green/turquoise blue. $25.00. [$15-20] Spring/Summer 1964.

Two-piece dress and sweater set. Yellow dress of Irish linen with cap sleeves, washable belt. Hand-embroidered wool sweater in yellow, blue, lilac, and pink on white. Dress is American-made, sweater imported from Hong Kong. $26.00. [$65-75] Spring/Summer 1964.

Shirtwaist dress in silk pongee imported from Hong Kong, hand-knotted buttons, hand-bound buttonholes, hand-stitched belt eyelets. Bright yellow. $23.50. [$15-20] Spring/Summer 1964.

Silk pongee sleeveless sheath imported from Hong Kong, fully lined, contrasting color on reverse of sash, and hand-made oriental ornament on pocket. Light beige with navy blue. $21.50. [$20-25] Spring/Summer 1964.

The overblouse dress in silk shantung, fully lined, hand-bound pocket with handmade frog ornament. Light aqua green. $29.50. [$25-30] Spring/Summer 1964.

Hand-screened garden print on silk surah, fully lined. Turquoise blue/royal blue on white. $21.50. [$35-40] Spring/Summer 1964.

23

The tweed look in a white-flecked nubby boucle knit with a frosty effect. Green. $17.50. [$35-40] Spring/Summer 1964.

Seersucker look in a new crinkled stripe boucle knit pullover shift. Orlon acrylic/nylon. Muted green and white stripes. $14.50. [$35-40] Spring/Summer 1964.

Splashy floral print dress with pleated skirt and contrast belt. Dacron® polyester crepe. Cool blue/green/white. $15.84. [$15-20] Spring/Summer 1964.

Bow-accented dress with unpressed box-pleated skirts. Belt with plastic bow trim. Dacron polyester crepe. Light aqua. $11.84. [$40-45] Spring/Summer 1964.

"City-bred Sheath" with the look of two pieces. Skirt has hidden pockets. Tweed-textured rayon and silk. Aqua. $14.84. [$35-40] Spring/Summer 1964.

Gold and Silver Color Embroidered Trim Shift. Boucle knit of Orlon acrylic and nylon. White. $16.50. [$35-40] **Middy Look 2-piece Dress.** White pebbly effect with navy crinkled striped trim. Boucle knit of Orlon acrylic and nylon. $22.50. [$35-40] Spring/Summer 1964.

24

From left to right:
Black jacketed dress of acetate and rayon crepe, black and white print acetate taffeta lining in jacket, sleeveless sheath. $10.84. [$20-25] Spring/Summer 1964.

Black and white optical print full skirted dress in Antron® nylon jersey. Cap sleeves. Black pompon-fringed acetate satin belt. $14.84. [$35-40] Spring/Summer 1964.

Ivory white textured cotton knit sheath, acetate taffeta lined, cord belt with decorative tassels. $13.84. [$15-20] Spring/Summer 1964.

Step-in-style coachman dress, in Arnel® triacetate and cotton with braid trim, mock pocket flaps, side pleats. Gray with black trim. $10.84. [$20-25] Spring/Summer 1964.

Cotton schiffli-embroidered scroll design shift in linen-look rayon. Optional cotton broadcloth belt. Gold on beige. $11.84. [$15-20] Spring/Summer 1964.

Garden Flower Print Jacketed Dress. Linen-look rayon jacket, Arnel® triacetate jersey dress with permanent crystal pleats. Yellow/aqua/white. $14.84. [$15-20]
Button-trimmed Bolero Jacketed Dress. Cotton/mohair/nylon jacket, acetate lined. Linen-look rayon sheath with V-shaped seamed front. Lemon yellow. $15.84. [$30-35] Spring/Summer 1964.

Black and white shift in acetate surah print, with optional self-belt. Sleeveless black linen-look rayon duster included. $12.54. [$40-45] Spring/Summer 1964.

Layered look 3-piece ensemble includes a button-back jerkin with mock pocket flaps and skirt in black rayon/silk, and a white blouse of acetate/rayon crepe. $19.84. [$20-25] Spring/Summer 1964.

25

Polka-dot Two-piece Dress. Cotton/rayon white polka dot print on pink. Overblouse has cowl neck that ties in back. $9.84. [$15-20] **Crystal Pleated Dress.** Arnel® triacetate jersey. Elasticized waist. Pink. $11.84. [$20-25] **Silk Surah Print Shift.** Fully lined in acetate taffeta, self belt. Pink and brown. $13.84. [$30-35] Spring/Summer 1964.

Flattering fashions in pink.
Suit-dress. Rayon and silk shantung, fully lined with acetate taffeta. $14.84.
Drawstring Bow Sheath. Softly shirred bodice with drawstring neckline. $13.84.
Rayon and Silk Shift. V-neck, fully lined in acetate taffeta. Dusty pink. $9.84. Spring/Summer 1964.

This Sears dress is "styled in the right size for you." Step-in style with a loop-over tie, pleated skirt. Arnel® triacetate and polyester crepe blend. Available in Juniors, Misses, Shorter Women, or Tall Misses sizes. Aqua and white. $17.50. [$15-20] Spring/Summer 1964.

Jacketed Dress styled in four sizes: Petite Juniors, Juniors, Misses, and Shorter Women. Slender back-zip sheath and bound-to-match jacket. Double-knit Arnel® triacetate jersey. Gold dress with gold/white print jacket. $21.50. [$30-35] Spring/Summer 1964.

Pure Irish linen dresses styled in four sizes: Juniors, Misses, Shorter Women, and Tall Misses. **Boat-neck Dress.** Two front pockets, back zip. Kelly Green. $14.50. [$25-30] **V-neck Dress.** Step-in style with covered buttons and decorative stitching around sleeves. Rayon chiffon scarf included. Pink. $18.50. [$15-20] Spring/Summer 1964.

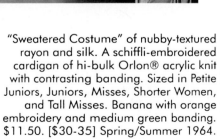

"Sweatered Costume" of nubby-textured rayon and silk. A schiffli-embroidered cardigan of hi-bulk Orlon® acrylic knit with contrasting banding. Sized in Petite Juniors, Juniors, Misses, Shorter Women, and Tall Misses. Banana with orange embroidery and medium green banding. $11.50. [$30-35] Spring/Summer 1964.

Mary Lewis designed coat-style shirtdress in polyester/combed cotton blend. Novelty tucks and self-piping on bodice front, billowy cuffed sleeves. In four sizes: Juniors, Misses, Shorter Women, and Tall Misses. Lilac. $11.50. [$15-20] Spring/Summer 1964.

27

Left:
All dresses in acetate and rayon crepe. **Smocked Dress.** Wear belted or loose. Aqua blue. $10.84. [$15-20] **Three-piece Outfit.** Sleeveless jacket lined with acetate taffeta. Blouse with collar tie and back buttons. Light rose pink with white. $11.84. [$15-20] **Square Neckline Sheath.** Rhinestone ornament at waist, fully lined in acetate taffeta. $11.84. [$30-35] **Classic Pleated Dress.** Jewel neckline, acetate-lined bodice, cluster box pleated skirt. Pin not included. Mint green. $9.84. [$15-20] Spring/Summer 1964.

Two-piece Dress. Rayon and acetate. Overblouse fully lined in acetate taffeta, self-covered buttons down back. Skirt suspends from back-zippered camisole. Lilac. $10.84. [$20-25] **Empire Shift.** Drawstring tie at high waistline. Acetate/rayon, fully lined in acetate taffeta. Hot pink. $8.84. [$30-35] Spring/Summer 1964.

28

Matte Jersey Sheath in acetate/nylon "feels like velvet", fully lined in acetate taffeta. Pin not included. Copen blue. $9.84. [$35-40] Spring/Summer 1964.

Three-piece dress of Arnel® triacetate jersey, ribbed white overblouse and skirt, lime green and turquoise striped jacket. Skirt fully lined in acetate taffeta. Available in two skirt lengths. Floor length, $20.84. [$25-30] Street length, $18.84. [$15-20] Spring/Summer 1964.

Printed Arnel® triacetate jersey sheath with self braid trim on bodice. Gold and white. $9.84. [$15-20] Spring/Summer 1964.

All dresses in solid color, floral print, polka-dot, and striped designs are Arnel® triacetate jersey. The slim shift styles are fully lined in acetate. $9.84-$11.84. [$20-25] Spring/Summer 1964.

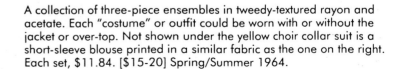

The "new" fabrics in crunchy weaves and textured look. **Double Button Coatdress.** Rayon/cotton/silk blend with "smoke ring" printed collar. Raspberry. $11.84. [$15-20] **Shift Dress with Paisley Jacket.** Rayon/acetate dress is bound with matching print. Sleeveless jacket fully lined in acetate taffeta. Beige with brown. $11.84. [$20-25] **Two-piece Rayon and Acetate Dress.** Zipper closed placket front with white braid trim. Turquoise and white. $8.84. [$15-20] Spring/Summer 1964.

A collection of three-piece ensembles in tweedy-textured rayon and acetate. Each "costume" or outfit could be worn with or without the jacket or over-top. Not shown under the yellow choir collar suit is a short-sleeve blouse printed in a similar fabric as the one on the right. Each set, $11.84. [$15-20] Spring/Summer 1964.

29

The coachman dress, a popular style, is shown here in the princess shape (left) and the eased shift (right) with two different collar treatments. Arnel® triacetate and cotton denim. Gold with gold and white polka-dot scarf included, and copen blue. Each, $10.54. [$15-20] Spring/Summer 1964.

The shift, another popular style, is shown in printed combed cotton sateen. The jewel neck shift in pink, light olive green, and aqua is completely lined in cotton broadcloth, and comes with an optional belt. The black and white boat neck shift features a "match-box" pleat at each side, and is outlined with braid trim. Fully lined in cotton with a braid belt included. Each, $8.84. [$30-35] Spring/Summer 1964.

The Four-way Shift. Can be worn with pull-through sash as shown, with the sash tied in back for a smooth front effect, unbelted, or belted all around. Linen-look rayon, lined in acetate. Hot pink on beige. $8.84. [$15-20] **Three-tone Three Piece Outfit.** Jacket and overblouse piped to match skirt. Cream, turquoise, and taupe. $12.54. [$15-20] Spring/Summer 1964.

The new back-wrap design in a two-piece slim style, and a swingy one piece dress. Clearly meant to be seen while walking, each dress is faced with a contrasting red polka-dot or flower print calico fabric. Light gray with red, and light olive green with red. Each, $9.84. [$30-35] Spring/Summer 1964.

This embroidered dress has a trapunto-style schiffli for a quilted look. Combed cotton lawn. White with pink trim. $9.54. [$35-40] Spring/Summer 1964.

Cool sheer fabrics and schiffli embroidery accent these summer dresses. In pima cotton or Dacron® polyester batiste. Shown in three fitted waist styles. $9.84-$11.84. [$35-45] Spring/Summer 1964.

Shirtdress with Tucks and Lace. Step-in style of sheer pima cotton. Light lilac. $10.84. [$20-25] **Pleated Dress.** Easy care Dacron® polyester batiste. Bow at back neckline. Medium pink. $12.54. [$25-30] Spring/Summer 1964.

A collection of ensembles and dresses in linen-look rayon. Note the popularity of schiffli-embroidery as it appears once again in the hot orange Spanish-style dress. $9.84-$14.84. [$35-55] Spring/Summer 1964.

31

Dresses in four different looks: gingham checks with inverted front pleats, straight line plaid shift, pullover one-piece pleated toile-look print, and a flared skirt skimmer. In cotton, polyester batiste, or rayon. $9.84-$10.84. [$25-35] Spring/Summer 1964.

Junior dresses in cotton knits and linen-look rayon. The red sleeveless coat two-piecer features the "new" saucer-shaped buttons. $9.84-$14.84. [$20-25] Spring/Summer 1964.

Summer dresses in various styles and fabric, in solid shades or printed florals. Note the four very different necklines. *From left.* In acetate and rayon crepe, polyester crepe, polyester batiste, and silk. $8.84-$11.84. [$20-25] Spring/Summer 1964.

An assortment of Junior fashions. $5.84-$7.84. [$25-30] Spring/Summer 1964.

An assortment of Junior fashions. Schiffli-embroidery accents the white and pink dress, second from right. Note that hemlines fall just at the knee. $5.84-$7.84. [$25-30] Spring/Summer 1964.

"Feel free, look crisp in carefree seersucker." In culotte, two-piece suitdress, and coatdress styles. The green outfit at right is cotton and acetate woven plaid seersucker. $5.50-$8.50. [$15-20] Spring/Summer 1964.

Novelty-knit Cardigan Coat. 7/8 length raglan-style sleeves, wooden buttons. Beige. $11.90. [$25-30] **Three-piece Double Knit Outfit.** Multicolor paisley print jacket. $19.90. [$15-20] **Two-piece Textured Knit.** Scarf not included. Geranium rose. $12.90. [$15-20] Spring/Summer 1964.

33

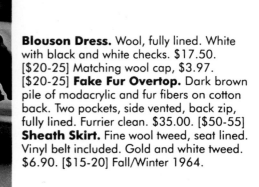

Blouson Dress. Wool, fully lined. White with black and white checks. $17.50. [$20-25] Matching wool cap, $3.97. [$20-25] **Fake Fur Overtop.** Dark brown pile of modacrylic and fur fibers on cotton back. Two pockets, side vented, back zip, fully lined. Furrier clean. $35.00. [$50-55] **Sheath Skirt.** Fine wool tweed, seat lined. Vinyl belt included. Gold and white tweed. $6.90. [$15-20] Fall/Winter 1964.

Boxy Cardigan Two-piece Set. White novelty banding. Lightweight, high bulk virgin Orlon® acrylic. Lilac. $8.97 [$15-20]. **Floral Pullover Shift.** Wear as is or belted with tie belt. Pink and turquoise on cream. $6.97. [$20-25] Spring/Summer 1964.

Straight shaft dress in deeply tweeded mohair and wool by *Einiger Mills*. Fully lined. Bittersweet blue and gold multicolor. $25.00. [$20-25] Fall/Winter 1964.

Dramatic Dot Shift. Balloon sleeves, self belt sash. Firmly woven imported cotton. Maple tan with black. $9.84. [$35-40] **Floral Pattern Ruffly Shift.** Empire bodice with sash belt. Acetate and nylon lace, fully lined. Black. $11.84. [$50-55] **Empire Stripe Shift.** Woven striped acetate and cotton seersucker. Blue with black. $5.54. [$40-45] Fall/Winter 1964.

Black and White Houndstooth Wool Jumper. Dickey length blousette of white acetate crepe and pink textured cotton and rayon tie included. $17.84. [$30-35] **Pink Two-piece Textured Weave Wool.** Reversible scarf, scalloped overblouse is bodice lined, skirt lined through hips. $15.84. [$15-20] **Turtleneck Sheath.** Wool jersey with cotton knit turtleneck. Pin not included. Light beige. $11.84. [$20-25] **Pleated Flannel Wool Dress.** Buttons trim waist and sleeves. Pin not included. Lilac. $12.54. [$15-20] Fall/Winter 1964.

More couture fashions by *Claude Riviere*. **Princess Slim Wool Dress.** Textured wool, fully lined. Aqua blue. $33.00. [$45-50] **Two-piece Textured Wool Dress.** Jacket has ¾ length sleeves. White. $39.00. [$35-40] **Glittering Princess Dress.** Gossamer-weight imported fabric of nylon, gold-color Mylar® metallic and rayon. Fully lined. Pink. $37.00. [$50-55] Fall/Winter 1964.

Empire-style Fitted Dress. Cotton velveteen, wear with or without sash. Red. $12.54. [$50-55] **Black and White Check Shift.** Pullover style, wear with or without belt (included). $7.84. [$25-30] **Batik-type Printed Dress.** Empire surplice bodice. Brown, blue, green, and beige cotton broadcloth. $6.84. [$40-45] Fall/Winter 1964.

35

White Blouson Dress. Basketweave wool, fully lined, pink bow, and cuffs of cotton and rayon shantung. $14.84. [$20-25] **Ruffle Trim Dress.** Airy weave, textured wool crepe, fully lined. Bright rose. $11.84. [$15-20] **Tailored Coachman Dress.** Step-in style, mock pocket flaps, back belt. Wool flannel. Scarf not included. Bright green. $11.84. [$15-20] **Wool Flannel Suit Dress.** Jacket piped in contrasting red, tiny pockets. Scarf not included. Navy blue with red trim. $12.54. [$15-20] Fall/Winter 1964.

Economically priced dresses in brushed rayon and acetate flannel. **Aqua Blue Sheath.** Rayon satin bow accent. Dress fabric has satin-weave back. $5.84. [$30-35] **Green and White Jumper Outfit.** Drop waist jumper with inverted front pleat. Dickey length blousette of printed striped acetate jersey. $7.84. [$15-20] **Camel Tan Suit Dress.** Jacket edged with contrasting brown braid trim. $5.84. [$20-25] Fall/Winter 1964.

Fashions in wool or Orlon® and wool, priced at just $8.00 each. [$15-20] *From left.* Box-pleated dress, Plaid sheath, Zip-front shift, Overblouse dress. Fall/Winter 1964.

Double-knit cotton dresses are shown in the latest fall colors. **Two-piece Dress.** Back-buttoned overblouse with pin. Price includes a 2-cent Federal Excise Tax on pin. Gold. $7.84. [$20-25] **Belted Shift.** Also worn without belt. Lantern red. $7.84. [$15-20] Fall/Winter 1964.

More fashions in wool or Orlon® and wool blend, economically priced at $8.00. [$15-20] *From left.* Blue coatdress, Red jersey dress (pin not included), Paisley sheath, Green wool flannel sheath dress. Fall/Winter 1964.

$7.84 [$20-25] buys so much fashion at Sears. **Stripe Three-piece.** Vestee and skirt of Dacron® polyester and cotton poplin. Cotton blouse with attached ascot. Dark green with rust, blue, green, and gold. **Shirtwaist Dress.** Embroidered mock monogram. Arnel® triacetate jersey. Light moss green. **Posy-printed Dress.** Acetate jersey with novelty belt. Brown, gold, and beige. **Tie-accented Two-piece Dress.** Nubby-textured rayon. Taupe brown. Fall/Winter 1964.

Two-piece tweed vest and A-line skirt of wool and nylon. Buttons and piping in soft vinyl. Heathery mix of moss green, black, and white. Shown here with a bow-tie blouse, not included. Vest and skirt set, $12.90. [$20-25] Fall/Winter 1964.

These wool jersey dresses were comfortable to wear because of the "intimate apparel" fabric worn next to the skin. SAG-NO-MOR® PLUS wool jersey, bonded to Certifab® acetate tricot. Gold, Bright green, and Bright navy blue. Each, $14.50. [$20-25] Fall/Winter 1964.

The classic shirtwaist is presented here with a shirred bodice front and a style with novelty pockets. In red and gold rayon and cotton broadcloth. Each, $5.00. [$15-20] Fall/Winter 1964.

Junior fashions in black and white. **Cotton Pique Swingy Skirt Dress.** Cord-rimmed at waistline and hem. $10.84. [$20-25] **Bowknot Pin.** Black and white polka dot enamel on metal. Approximately 3" wide. $1.00. **Two-piece Belted Dress.** Pebble-textured double-knit cotton with cowl neck overblouse and slip-through belt. $13.54. [$35-40] **Three-piece Outfit.** Woven striped Dacron® polyester and cotton seersucker jacket and skirt. Sleeveless blouse is polyester and cotton voile with elasticized waist. Set, $17.84. [$30-35] Spring/Summer 1965.

Two-piece jacket and dress is in linen-look rayon. Jacket is tucked in front and white frosted with corsage and curving collar. Black and white. $12.84. [$50-55] Spring/Summer 1965.

This two-piece ensemble was worn as a combination dress and light coat, or separately when each part is worn as a dress. Lace-rimmed coat of American flock-dotted Swiss in polyester and rayon, cotton lined except for sleeves. Dress is linen-look rayon. Back and white. $20.84. [$40-50] Spring/Summer 1965.

Double duty two-piece outfits could be worn as a set or separately. **Black and White Houndstooth Check Duster and Yellow Dress.** Duster is woven cotton trimmed with rayon braid, fully lined. Yellow dress is linen-look rayon. There is no mention of the hat. $21.84. **Orange Sleeveless Duster and Print Dress.** Outfit of rayon and linen. Nubby textured duster has front pocket and back belt. Fitted dress in orange, gold, and green floral print on beige. $15.84. Spring/Summer 1965.

Eyelet-embroidered Ruffles and Tucks Shirtwaist. Dacron® polyester and cotton broadcloth. Yellow. $10.84. [$35-40] **Embroidered Lattice Work Dress.** Dacron® polyester batiste. Peach. $10.84. [$30-35] **Schiffli-embroidery Striped Dress.** Sheer nylon organdy on bodice and overskirt, over woven striped cotton broadcloth dress. Embroidered flowers and scallop edging. Light blue and white. $14.84. [$30-35] Spring/Summer 1965.

39

Pullover Dress. Linen-look rayon dress, white collar, braid and button opening. Back belt. Bright green and white. $6.84. **Scallop Collar Skimmer.** White embroidered trim and bud. Bright navy and white. $8.84.

Polka-dot Ruffles. Rayon and cotton, fully lined, plastic patent belt. Black and white. $8.84. [$25-30] **Ruffle Top Dress.** One piece dress with a two-piece look. Bodice of Dacron® polyester and cotton voile, skirt of woven striped acetate and cotton cord. Light red with red and white stripes. $8.84. [$30-35] Spring/Summer 1965.

Colorful Stripe Pleated Dress. Linen-look rayon in rose, green, yellow, orange, blue and white. $6.84. [$15-20] **Wide Collar Dress.** Dacron® polyester and cotton voile. Plastic belt included. Dark spruce green and white. $9.84. [$15-20] **Woven Check Blouson Dress.** Ruffle trim at neckline. Back buttons to waist, and zips below waist. $6.84. [$20-25] Spring/Summer 1965.

Checkered Jacket Dress. Bodice of dress in woven check cotton gingham. Skirt and bolero is rayon and cotton poplin. Bolero lined in matching gingham. Brown with turquoise blue and white checks. $7.84. [$30-35] **Large Gingham Check Shirtdress.** Button-on kerchief-collar detaches to wear as a scarf. Yellow and white. $6.84. [$30-35] Spring/Summer 1965.

Sailor Skimmer. Zip-front with mock pockets. Linen-look rayon. White with navy. $6.84. [$30-35] **Two-piece Middy Dress.** Navy linen-look rayon pullover with red and white striped snap-in insert. Skirt in pleated Arnel® triacetate jersey. $8.84. [$25-30] Spring/Summer 1965.

Dots and Stripes Set. Dress has cotton bodice in a dot pattern. Skirt and jacket is linen-look rayon stripes. Jacket lining matches bodice. Navy and white. $12.84. [$20-25] **Shaped Dress with Embroidery Trim.** Nubby-textured linen-look rayon dress with braid embroidery scalloped collar. White with navy. $13.54. [$30-35] Spring/Summer 1965.

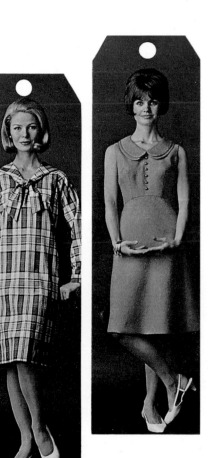

Pleated Two-piece Dress. Arnel® triacetate and polyester crepe dress with nylon organdy bow. Navy and white print. $16.84. [$15-20] **Two-piece Suit Dress.** Silk shantung, fully lined in acetate taffeta. Navy blue. $19.84. [$15-20] **Polka-dot Pouf Hat.** Surah of rayon and silk, lined in rayon organza. Fluff it up, down, or to the side. White with navy dots. $5.97. [$25-30] Spring/Summer 1965.

Middy Shift. Woven cotton plaid in blue, gray, black, and red. Pullover style with sailor collar. Sash to wear belted is included, not shown. $5.84. [$15-20] **Double Collar Dress.** Textured rayon and silk, fully lined. Orange. $9.84. [$15-20] Spring/Summer 1965.

A two-piece jacket dress with a three-piece look. Blouse bodice of filmy polyester chiffon. Skirt and jacket in polyester crepe. Both lined. Light pink. $25.00. [$30-35] Spring/Summer 1965.

Bow-trimmed Dress. Linen-look rayon, fully lined. Light pink. $11.84. [$40-45] **Slender Dress.** Acetate and rayon crepe, lined in acetate taffeta. Gathered at yoke, darted at waistline. Light pink. $11.84. [$20-25] **Three-piece Outfit.** Rayon and linen. Cardigan jacket has diagonally placed contrasting color bound buttonholes, with facing that matches blouse. Rose pink and light pink blouse and trim. $13.84. [$35-40] Spring/Summer 1965.

Soft floral dress of printed rayon chiffon, fully lined in acetate taffeta. Bodice edged with self-piping. Acetate satin tie-belt. Rose, orchid, green, and pink. $15.84. [$40-45] Spring/Summer 1965.

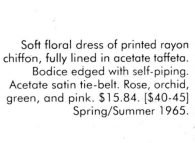

Sears featured a section on "travel-wise clothes you can pack into one bag." **Printed Shift.** Pure silk jersey in red, gold, and raspberry. $15.84. [$30-35] **Two-piece Knit Dress.** Orlon® acrylic and nylon knit. Gold and white. $17.84. [$20-25] **Three-piece Plaid Suit Dress.** Jacket and skirt in woven multicolor plaid seersucker of cotton and acetate. Not seen is a polyester and cotton broadcloth sleeveless blouse that buttons in back. Brown/gold/black on blue plaid, Copen blue blouse. $12.84. [$15-20] Spring/Summer 1965.

More travel-wise clothing that features easy care and easy packing. **Pleated Skirt Dress.** Dacron® polyester and cotton broadcloth. Navy blue with light gold-color trim. $12.84. [$25-30] **Two-piece Jacket Dress.** White cotton twill with gold-color buttons and chain belt. $13.84. [$15-20] **Coat and Dress Ensemble.** Woven check trenchcoat of rayon, acetate and linen, lined with acetate taffeta. Rayon and acetate short-sleeve sheath dress, edged at neckline with matching check fabric. Light blue and white checks, light blue dress. $22.84. [$20-25] Spring/Summer 1965.

Anytime Trio. White rib-textured shell and skirt with an orange and yellow printed jacket with the feel of silk. Arnel® triacetate jersey. Jacket is fully lined in acetate taffeta. Floor length, $22.84. [$30-35] Street length, $20.84. [$25-30] Spring/Summer 1965.

Citrus shades were the "hot" colors for spring. **Two-piece Polka-dot Ensemble.** Shift dress of acetate surah, sleeveless coat of textured rayon and silk. Lemon and white. $12.84. [$25-30] **Blouson Dress.** Chelsea collar. Grainy textured printed Dacron® polyester crepe, fully lined. Lemon, lime, and olive on white. $13.84. [$18-20] **Shirtdress.** Gossamer rayon chiffon with attached acetate taffeta slip. Stitched tucks trim bodice front, collar, and cuffs. Lemon. $14.84. [$15-20] **Twin-Print Two-piece Dress.** Cowl neck rayon chiffon blouse with elastic waist, lined except for sleeves. Skirt is linen-look rayon. Lemon, orange, lime, and turquoise floral print. $13.84. [$20-25] **Scarf Dress.** Two-piece textured rayon and silk, fully lined in acetate taffeta. Tri-tone printed rayon chiffon "smoke ring" scarf included. Spring/Summer 1965.

Floral Ruffle Scoop Neck Shift. Silk-like Arnel® triacetate jersey in a multicolor screen print, fully lined. Pullover style, sash belt included. $10.84. [$25-30] **Floral Blouson Dress.** Fitted underbodice, button-back blouson bodice. Arnel® triacetate jersey, fully lined. Orange, coral, gray, and old gold print on white. $11.84. [$25-30] Spring/Summer 1965.

Upholstery Print Dress. Heavy textured cotton. Skirt is lined. Multicolor tones. $11.84. [$25-30] **Three-piece Suit Dress.** Jacket and skirt of rayon twill, fully lined in acetate. Screen-printed jersey pullover blouse of Arnel® triacetate nylon. Orange with orange, pink, and white blouse. $19.84. [$20-25] Spring/Summer 1965.

44

Sears offered these four dress styles sized to fit Petite Juniors, Juniors, Misses, and Tall Misses. Each dress was priced at $7.84. [$15-25] **The Two-piece Dress.** Textured double-knit cotton. Bright green. **The Shirtwaist.** Dacron® polyester and cotton oxford cloth. Coat-styled and double collared, novelty belt. Blue. **The Sheath.** Rayon and acetate crepe, fully lined in acetate taffeta. Pin not included. Pink. **The Three-piece Suit Dress.** Jacket and skirt in Dacron® polyester and cotton poplin. Flower printed button-back cotton broadcloth overblouse. Yellow. Spring/Summer 1965.

Butterfly Shirtdress. American flock-dotted Swiss of Dacron® polyester and cotton. Concealed buttons under applique embroidered butterflies. Pink with white. $6.84. [$15-20] **Eyelet Coachman Shift.** Flower patterned eyelet-embroidered acetate and nylon. Brass-color buttons on front and on back belt. White. $6.84. [$15-20] Spring/Summer 1965.

Classic cotton shirtwaist in three patterns were available in four sizes: Junior, Miss, Shorter Women, and Petite Junior. Each dress comes with a hemp belt. Shown here in woven tattersall-checked gingham, floral printed broadcloth, and woven striped seersucker. Priced at an affordable $5.84 each. [$15-20] Spring/Summer 1965.

45

Pretty dresses for only $5.00. [$15-25] **Chelsea-collared Dress.** American flock-dotted Swiss of cotton. Collar and neckline insert of linen-look rayon. Navy blue with white. **Abstract Print Dress.** Rayon and cotton crepe, plastic rope belt. Aqua, sand, beige, and green print. Spring/Summer 1965.

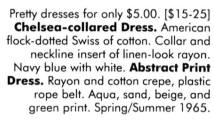

Polka-dot Sweater Dress. Rayon and cotton step-in dress, cotton knit sweater. Turquoise with white dots. $5.00. [$40-45] **Woven Stripe Two-piece Dress.** Acetate and cotton seersucker. Mock pockets. Black and white. $5.00. [$18-20] **Tucked Bodice Floral Dress.** Polyester and rayon broadcloth. Light and dark lavender, and gray on white. $5.00. [$15-20] Spring/Summer 1965.

The Schiffli-embroidered cotton broadcloth dress came in full-skirted or slim styles. The full-skirted style has short sleeves with turn-back cuffs. The slim-skirted version featured ¾ length sleeves, barrel cuffs, and two pockets. Pink. $5.84 each. [$20-25] Spring/Summer 1965.

Breeze-weight sheers in Dacron® polyester batiste. **Schiffli-embroidered Dress.** White flowers and cotton lace on bodice front. Salmon pink. $9.84. **Muted floral Dress.** Self piping detail, button-trimmed bodice. Pink, light coral, and beige on white. $8.84. Spring/Summer 1965.

Elegant slim dresses with the look of linen. **Coachman-style Dress.** Diagonal weave rayon and silk. Brass-color buttons are decorative only, dress zips in back. Light beige. $8.84. [$15-20] **Collared Suit Dress.** Two-piece rayon dress. Jacket with self covered buttons, faced in acetate taffeta. Caribbean blue. $9.84. [$20-25] Spring/Summer 1965.

These dresses received the Sears Four Star Value stamp for its high quality and low price. Made of Arnel® triacetate jersey, the dresses featured zippers enameled to match, and a low price of $9.84. Spring/Summer 1965.

Cool and light dresses in Dacron® polyester Whipped Cream® crepe. **Ruffly Woven Plaid Dress.** Rope style tie-belt. Apple green on oyster white. $10.54. [$20-25] **Blossoms Dress.** Bodice lined with acetate taffeta. Yellow and lime green on white. $10.54. [$20-25] Spring/Summer 1965.

Lightweight denim in a soft blend of Arnel® triacetate and cotton. **Bowed Two-piece Dress.** Trimmed in rayon grosgrain braid. Pullover top with mock pockets. Gray with black. $8.84. [$30-35] **Military Style Shift.** Trimmed in rayon grosgrain braid. Detachable white linen-look rayon collar. Gray with black. $9.84. [$25-30] Spring/Summer 1965.

47

Budget-priced dresses in easy-care Dacron® polyester and cotton broadcloth. **Ombre-shaded Plaid Dress.** Blue, green, and beige. $5.00. [$15-20] **Printed Stripe Shirtdress.** Open V-neck style adds a unique variation to the shirtdress. Sky blue, tan, lavender, and white. $5.00. [$15-20] Spring/Summer 1965.

Pure linen coordinates in a hand-crocheted look intricate knit. Rice white with olive green trim. Cardigan, $7.97. [$35-40] Shell, $4.97. [$25-30] Skirt, $5.97. [$25-30] Spring/Summer 1965.

"Young ideas for Saturday night." **Chelsea.** Acetate and rayon crepe, lined. Light beige. $11.84. [$15-20] **Lacy Middy.** Two-piece lacy knit wool dress, lined except for sleeves. Off-white. $19.84. [$20-25] **Princess Dress.** A-line dress of loopy wool and nylon, lined except for sleeves. Coral. $17.84.[$20-25] **Point D'esprit Patterned Stretch Nylons.** Seamed. Shown in white. $2.57 pair. [NPA] Fall/Winter 1965.

Junior fashions featuring the popular madras look. All plaids in these dresses are American madras-look woven cotton. They are machine washable and will not bleed like authentic madras cloth. **Skimmer.** Sage green skirt and matching necktie in cotton poplin. $6.84. [$20-25] **Plaid Shift.** Step-in style dress with self sash and red tie included. $5.84. [$15-20] **Drop-waist Dress.** Plaid bodice, navy blue cotton poplin skirt, yellow plastic belt. $6.84. [$20-25] Fall/Winter 1965.

Sporty Two-piece "T" Dress. Double-knit cotton. Red and gold striped overblouse with low-slung belt. $13.84. [$20-25] **Two-piece Hip Rider Dress.** Red cotton knit pullover top. Red, coral, and gray woven checked acrylic and rayon. $9.84. [$25-30] Fall/Winter 1965.

Overblouse Dress. Two-piece silk shantung, fully lined in acetate. Turquoise blue. $17.84. [$20-25] **Flanged Shoulderline Sheath.** Bowed self belt. Silk shantung, fully lined. Pin not included. Medium beige. $15.84. [$20-25] **Roll Collar Suit Dress.** Rhinestone buttons and loop fasteners. Silk shantung, fully lined. Moss green. $19.84. [$20-25] Fall/Winter 1965.

Turtleneck Suit Dress. Jacket and skirt of loopy wool, mohair, and nylon. Jacket bodice lined in acetate. Snap-in white wool dickey hooks in back. Jade green with white. $19.84. [$20-25] Fall/Winter 1965.

Crochet Front Suit Dress. Two-piece double-knit wool with crochet and tucking trim down front and on cuffs. Crochet-covered buttons. Made in Italy. Camel brown. $28.50. [$20-25] **Contrasting Trim Three-piece Dress.** Three-piece double-knit wool made in Italy. Jacket with contrasting trim and two pockets. Rose red with black. $34.50. [$20-25] **Tweed-look Dress.** Pullover top has suede trim at neckline and pockets. Lamb's wool from Australia. Light willow green with dark green trim. $22.50. [$20-25] Fall/Winter 1965.

Cool sea shades of blue, turquoise, green, and white in a variety of summer dresses. **Two-Piece Diamond Plaid Dress.** Cowl collar ties in back. Printed polyester chiffon. $29.00. [$15-20] **Double-knit Sheath.** Ribbed polyester with self-belted included. Pin not included. $19.90. [$15-20] **Silk Skimmer.** Floral-printed silk, fully lined in acetate. $14.50. [$25-30] **A-line Knit Dress.** Cotton knit with high rib-knit cowl collar. Slip-through black plastic belt included. $10.90. [$25-30] **Polka-dot Suit Dress.** Rayon and silk that looks like linen. Jacket fully lined. $23.90. [$35-40] **Scarf Dress.** Two-piece linen-look rayon, sleeves and scarf of rayon chiffon. $12.50. [$30-35] Spring/Summer 1966.

Striking yellow and navy blue two-tone dresses are presented here in a one piece "skimp" dress, and a two-piece overblouse and skirt set. Acetate knit. Each, $14.75. [$40-45] Spring/Summer 1966.

Perfect for travel, these outfits made of Arnel® triacetate jersey are lightweight, and resist wrinkling. **Striped Jacketed Dress.** Double-knit sheath with leather-look plastic belt. Jacket edged with narrow braid. White with black. $19.90. [$15-20] **Houndstooth Print Sheath.** Front overblouse effect. Vertical border pattern is in a bolder print. Lined with acetate taffeta. Rayon grosgrain belt. Black and white. $12.90. [$15-20] Spring/Summer 1966.

Each of these great styles may be purchased for just $10.00. [$25-35] **Check Jacket Dress.** Jacket and skirt portion of dress in woven checked cotton and triacetate, bodice in cotton broadcloth. Bright sky blue and white check. **Two-piece Denim Dress.** Arnel® triacetate and cotton, plastic belt and buttons in contrasting white. Light gold. **Textured Rayon Three-piece Dress.** Rayon and cotton blend, striped blouse of polyester and rayon. Light celery green with multicolor striped blouse. **White Rimmed Ruffle Dress.** Polyester and cotton voile shift, fully lined except sleeves. Light burgundy red. Spring/Summer 1966.

51

The "new sunny look" of California in bold bright colors, and cool linen-look rayon. **Artful Skimmer.** Mondrian-inspired contemporary dress. Currant red and pink, bordered by white. $8.90. [$35-40] **Chelsea Skimmer.** Contrasting collar and front bodice. Kelly green with light beige. $9.90. [$35-40] **Jacket Dress.** Sheath with light beige bodice and light taupe brown skirt. Belt matches orange jacket. $9.90. [$15-20] **Striped Skimmer.** Bold panels of teal blue, aqua blue, and chartreuse. $7.90. [$20-25] Spring/Summer 1966.

More beautiful dresses, economically priced at $10.00 each. [$15-25] **Lacy Shift.** Textured rayon and cotton shift, with cotton lace sleeves. Light beige. **Drop Waist Shift.** Orlon® acrylic knit, bonded to acetate tricot. Wear belted at natural waist, or without. Pink. **Paisley Dress.** Very lightweight batiste polyester dress. Light jade green and pink on bright gold. **Striped Two-piece Dress.** Dacron® polyester and rayon in bright striped tones of blue and green. Top zips in front. Spring/Summer 1966.

Pleated Dress. Dacron® polyester and cotton poplin. Shirred bodice with smocking trim at waist. Copen blue with white trim. $9.50. [$20-25] **Chelsea Jacket Dress.** Dacron® polyester and cotton poplin. Sleeveless sheath with white top, solid color skirt. Jacket of printed stripes. Copen blue with white. $10.50. [$20-25] Spring/Summer 1966.

Loopy textured cotton knit fashions in bright green. Shown here in a skimmer with contrasting white trim, and a two-piece dress trimmed with a "carefree school-girl's tie." Each, $8.00. [$30-35] Spring/Summer 1966.

Sunny yellow dresses made of pure linen from Belgium. Shown in a roll-collar and tie top slender dress, and a two piece Chelsea-collared dress accented by a polka-dot bow. Each, $8.00. [$40-50] Spring/Summer 1966.

Calico-look Dress and Jacket. Printed cotton voile jacket and skirt-portion of dress. Bodice of dress, not shown, is yellow linen-look rayon. Yellow print on grass green. $17.90. [$30-35] **Little Girl's Jumper Shift.** Daisy print calico look cotton dress to wear as a jumper or shift. $4.90. [$18-25] Spring/Summer 1966.

53

Popular textured hand-crocheted look. Linen, acetate, and Orlon® acrylic knit. **Pullover Shift.** Open-knit lacy-look yoke and trim with optional self-belt. Pink. $8.97. [$15-20] **Short-sleeve Cardigan.** Open-knit lacy pattern with braid trim. Light beige. $7.97. [$10-12] **Matching Slim Skirt.** Rayon taffeta-lined. $6.97. [$10-12] **Pullover Scoop-neck Shell.** Contrasting crochet edging, bonbon studded front. Light beige. $5.97. [$20-25] **Matching Slim Skirt.** $5.97. [$10-12] Spring/Summer 1966.

Bold outlines were the "new" look for the season, fashioned in these Mondrian-inspired designs in navy blue and grass green. **Jacket.** Rayon hopsacking fully lined with acetate taffeta. Mock pockets, ¾ length sleeves. $10.97. [$12-15] **A-shaped Skirt.** Rayon hopsacking fully lined with acetate taffeta. Contour bandless waist. $6.97. [$8-10] Inset. **Sleeveless Shell.** White rayon hopsacking. $4.97. [$8-10] **Shift.** Rayon hopsacking fully lined with acetate taffeta. Pin not included. $9.97. **Overblouse.** Rayon hopsacking. Back button closing. $5.97. [$20-25] **Trumpet Pants.** Contour bandless waist. Rayon hopsacking fully lined with acetate taffeta. $8.97. [$30-35] Inset. **Scoop-brim Fedora.** Linen-look rayon with ribbon bow. Light beige. $4.97. [$30-35] Spring/Summer 1966.

Women's Fashions

Casual Dresses

"It costs little to look lovely." Assorted casual dresses in a variety of styles and fabrics. The pink dress features schiffli-embroidery in the front. These casual dresses are affordably priced from $5.84-$7.84. [$20-35] Spring/Summer 1964.

Floral and striped styles in "the modern broadcloth - lustrous and long wearing." In Dacron® polyester and combed cotton blend. Styled proportionately to fit four sizes: Juniors, Misses, Shorter Women, and Tall Misses. Only $6.84 each. [$20-25] Spring/Summer 1964.

Denim Jacket Dress. Top of dress is acetate and cotton woven striped seersucker with denim trim. Blue and white with red stitching. $7.84. [$15-20] **Pullover Plaid Dress.** Two-piece woven plaid of textured cotton and Arnel® triacetate. Pink and white. $6.84. [$15-20] Spring/Summer 1964.

The pink dress in printed cotton percale features a cotton knit sweater cuffed and edged in a matching fabric. Cherry pink and coral. $5.00. [$45-55] Spring/Summer 1964.

At Sears a woman may choose the style most flattering to her. These dresses were available in your choice of slim or full-skirted styles. **Schiffli-embroidered Cotton Broadcloth.** Convertible collar, embroidered front bodice, cuffed sleeves. Apricot. $5.84. [$30-35] **Madras-look Dan River American Cotton.** Collarless dress with elasticized straw belt. Blue, gray, and green woven plaid. $5.84. [$20-25] Spring/Summer 1964.

Polka-dot Print Step-in Dress. Rayon with linen-look collar. Navy blue and white. $5.00. **Applique Step-in Dress.** Dacron® polyester and cotton. Aqua blue. $5.00. Spring/Summer 1964.

The A-line blue and white woven cotton check shift with rick-rack trim can be worn with or without a belt. $3.84. [$30-35] Other cotton dresses are all nicely detailed, priced low, and machine washable. Each, $3.84. [$20-25] Spring/Summer 1964.

Touches of the
Old West make
fashion news today

Ruffle Jumper and Kerchief. Floral-printed cotton voile. Marigold and white. $12.90. [$40-45] **Lace Skimmer.** Cotton and polyester over cotton lawn lining. Lace ruffles trimmed with little white bows. White over yellow. $13.90. [$35-40] **Calico-Print Ruffle Skimmer.** Rayon and cotton broadcloth. Yellow, green, and white on black. $7.90. [$40-45] *Inset.* **Tote Bag.** Straw-like rayon imported from Japan. Bone beige, or White. $5.97. [$20-25] Spring/Summer 1966.

Button-front cotton shifts with convertible collar, accent stitching, two pockets, and slender belt. Shown in woven plaid, solid-color poplin, and woven check gingham. Each, $5.84. [$15-20] Fall/Winter 1965.

Women's Fashions

Casual Separates

Two-piece vest and skirt outfit in bright rose linen and cotton. Sleeveless vest detailed in covered buttons and mock pockets. $9.90. [$35-40] Spring/Summer 1964.

Alpaca sweater of subtle looped texture, V-neckline, turnback cuffs. Coral pink. $16.70. [$10-12] Spring/Summer 1964.

Interchangeable fashions in eyelet-embroidered cotton, each lined in cotton. Frosty white. Cropped top, $2.90. [$15-20] Long skirt with pink rayon sash, $8.90. [$25-30] Long sleeve overblouse, $5.90. [$15-20] Slim short skirt, $4.90. [$15-20] Ruffled top, $3.90. [$30-35] Capri pants, $4.90. [$35-45] Spring/Summer 1964.

Imported handmade mohair sweaters from Italy. ¾ length sleeves, in V-neck pullover or cardigan style. Each, $9.80. [$20-25] Spring/Summer 1964.

Lightweight jackets include a white cardigan in horizontal stretch cotton and nylon denim, and a red raglan-sleeve drawstring waist in stretch cotton duck. Each, $5.87. [$25-30] Spring/Summer 1964.

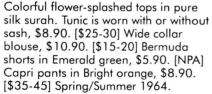

Colorful flower-splashed tops in pure silk surah. Tunic is worn with or without sash, $8.90. [$25-30] Wide collar blouse, $10.90. [$15-20] Bermuda shorts in Emerald green, $5.90. [NPA] Capri pants in Bright orange, $8.90. [$35-45] Spring/Summer 1964.

A collection of stretch wear fashions. Shift in horizontal stretch double-knit cotton/nylon, black and white stripe, $6.90. [$15-20] Navy blue walking shorts in stretch denim, $3.67. [$20-25] Checked Riviera style shorts in stretch nylon knit, $7.97. [$20-25] Capri set in two way stretch nylon/polyester. $14.97. [$40-45] Spring/Summer 1964.

"Bravo" fashions in hot bright colors. All in machine washable textured cotton. Hat in four colors, $2.47. [$20-25] The striped shirt and skirt outfit on left features a rose-red overskirt with attached walking shorts. Striped shirt, $2.87. [$20-25] Overskirt, $4.87. [$20-25] Striped shift, $4.87. [$15-20] Two-piece tango orange top and capri pants, $6.87. [$40-45] Senorita shirt with multicolor embroidery, $4.87. [$15-20] Wrap skirt, $3.87. [$12-15] Spring/Summer 1964.

Blue and white woven checks with lace and rick-rack trim. $2.87-$5.87. [$20-45] Spring/Summer 1964.

An assortment of skirt styles. **Front wrap.** Double coachman styling with contrasting buttons. Combed cotton denim. Faded blue. $3.97. [$12-15] **Two-way wrap.** Wear as front or back wrap. Polyester and cotton poplin. Brown. $4.97. [$12-15] **Sporty Skirt.** Pockets stitched in contrasting color. Combed cotton chino. Bright red. $4.97. [$15-20] **Hip-stitched Plaid Skirt.** Dacron® polyester and cotton woven plaid. $6.97. [$12-15] Spring/Summer 1964.

Jacket-blouses in textured natural linen look of acetate and cotton or multi-stripe Arnel® triacetate sharkskin. $4.87 and $5.87. Spring/Summer 1964.

60

Easy care cotton casual shirts to wear tucked in or out. Each, $2.87. [$10-12] Spring/Summer 1964.

Jacket Shirt in Arnel® triacetate surah print to wear in or out. $6.90. [$12-15] **Fisherman-style Pullover.** Acetate print that looks like silk. $4.90. [$12-15] Spring/Summer 1964.

16

17

Long-tail shirts in solid color or country print. Mandarin collar, roll sleeves. Each, $2.87. [$12-15] Spring/Summer 1964.

The pants look in traditional and new styles. **Front-wrap Culottes.** Red cotton duck. $3.87. [$15-20] **A-line Culottes.** Polyester and cotton poplin. Light green. $4.87. [$10-12] **Dutch Boy Pants.** Royal blue cotton duck. $2.87. [$20-25] **Cabin Boy Plaid Pants.** Woven cotton plaid with hemp belt. $3.87. [$15-20] **Pedal Pushers.** High-rise waist, plastic belt. Cotton gabardine. Hot pink. $2.87. [$20-25] Spring/Summer 1964.

Classic pants in four different lengths and 12 different colors. Shown here are Old gold, Lilac, Stone green, and pink. $2.87-$3.87. [$15-35] Spring/Summer 1964.

62

Cotton knit coordinates in rose pink and white. Rugby Shirt. $3.87. [$10-12] Stretch Pants. $5.87. [$10-12] Mock Monogram Overshirt. $2.87. [$10-12] Walking Shorts. $2.87. [$10-12] Sleeveless Pullover. $2.87. [$10-12] Cabin-boy Pants. $3.87. [$20-25] Sawtooth Hem Top. $2.87. [$20-25] Capri Pants. $3.87. [$15-20] Spring/Summer 1964.

Fortrel® polyester and cotton separates in medium green and sky blue. Belts reverse from check to solid and are included with pants and pert skirt (attached Bermuda shorts). Shirts, $2.87-$3.87. [$12-15] Shorts, skirts, and pants, $3.87-$4.87. [$15-20] Spring/Summer 1964.

Tops to compliment popular stretch pants. **Long Raglan Sweater.** Worsted wool knit. Winter white with heather gray tones. $26.90. [$50-55] **Poncho.** Under arm snap closing. Mohair, wool, and nylon. Heather gray, pistachio, and winter white. $15.90. [$35-40] **Side Zip Jacket.** Zip pocket on sleeve and at side, raglan sleeves. Matching triangular scarf included. Multicolor screen-printed cotton poplin, lined in nylon quilted to Dacron® polyester. $14.90. [$25-30] Fall/Winter 1964.

Fashionable sweaters with fine detailing. **Cardigan.** Brushed virgin Orlon® acrylic knit. Binding and button covering of sueded vinyl-coated fabric. Pastel pink with gray trim. $5.87. [$10-12] **Novelty Knit Jacket Sweater.** Orlon® acrylic, ¾ length sleeves. Beige. $6.87. [$10-12] **Floral Print Cardigan.** Virgin Orlon® acrylic knit. Lilac, green, and cream. $4.87. [$15-20] Spring/Summer 1964.

A collection of sweaters and perfectly matched separates shown in Deep pink, Blue, and Lilac. **Cardigan.** Wool and kid mohair. $6.84. [$12-15] **Slim Skirt.** Wool flannel, seat lined, walking back pleat. $4.84. [$12-15] **V-neck Pullover.** Wool and kid mohair. $5.84. **Tapered Pants.** Fully lined wool flannel. $6.84. [$12-15] **Sleeveless Pullover.** Wool and kid mohair. $4.84. [$15-20] **Knife-pleated Skirt.** Wool flannel. $5.84. [$12-15] Fall/Winter 1964.

Rainbow Pastel V-neck Pullover Sweater. Wool, mohair, and nylon blend. Hand knit from Italy. Pink, olive, and white. $10.90. [$15-20] **Hand Crocheted Cardigan.** Wool, mohair, and nylon in airy shell stitch pattern. White. $11.90. [$12-15] Fall/Winter 1964.

Sleeveless Pullover. V-neck, mock pockets, side slits. Red wool flannel. $5.90. [$15-20] **Ruffled Blouse.** V-neck, back button closing. White Arnel® triacetate crepe. $5.90. [$12-15] **Pleated Skirt.** Black and white houndstooth checked wool. $7.90. [$10-12] Fall/Winter 1964.

Two beautiful hand-knit and hand-stitched sweaters imported from Italy in a blend of wool, mohair, and nylon. In crew-neck cardigan and V-neck pullover styles. Pink, and Light green. Each, $9.90. [$12-15] Fall/Winter 1964.

Above left. "Only at Sears in America!" Brushed kid mohair, sheared from the youngest of the flock, were available to knit your own sweaters. These 90% kid mohair yarns were brushed to draw out the fibers, then fortified with 10% wool. The 1.4 oz. balls came in 15 colors. Shown in Light blue. $1.29 each. [NPA] *Right.* 100% mohair imported from Italy. In 14 colors. Shown in Beige. $1.09 each. [NPA] Fall/Winter 1964.

Jersey print blouses in three styles. **Turtleneck Abstract Print.** Turquoise blue and white. $3.84. [$15-20] **V-neck Tuck-in Blouse.** Pink and green rosebud print. $2.84. [$20-25] **Cowl-collared Overblouse.** Pink, beige, and maize tones on white. $3.84. [$15-20] Spring/Summer 1965.

Elegant and soft crepe blouses in a blend of acetate and nylon. Shown in mint green, white, maize, and blue. Each, $4.97. [$15-20] Spring/Summer 1965.

Belted style pocket shirts with contrast stitching, in combed cotton. Shown in woven plaid and solid color broadcloth. Each, $3.70. [$20-25] Spring/Summer 1965.

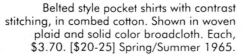

Dotted Swiss, an American cotton that has the dots woven right into the fabric, is shown here in coordinated red and white fashions. Pants, straight skirt, and jacket fully lined in cotton broadcloth. **Ruffled Tiered Top.** Cotton organdy with cowl collar, and back button closing. $3.84. [$30-35] **Capri Pants.** Fully lined. $4.84. [$40-45] **Pullover Top.** White with white dots, trimmed in red with white dots. $4.84. **Walking Shorts.** Fully lined. $3.84. [$20-25] **Chelsea-collared Pullover.** White tie of Arnel® triacetate sharkskin. Bodice lined. $4.84. [$30-35] **Pleated Skirt.** White Arnel® triacetate sharkskin. $4.84. [$15-20] **Jacket.** White braid edging, fully lined. $5.84. [$25-30] **Straight Skirt.** Fully lined. $3.84. [$15-20] Spring/Summer 1965.

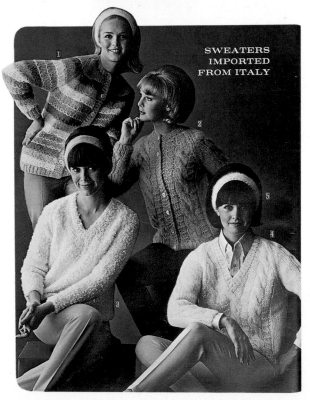

SWEATERS IMPORTED FROM ITALY

Right. White Embroidered Eyelet Fashions. Cotton birdseye pique n a diamond weave. **Sleeveless Shift.** Cotton broadcloth lined. $7.84. [$25-30] **Overblouse.** ¾ length sleeves, full length back zipper. $5.84. [$20-25] **Capri Pants.** $6.84. **Sleeveless Shell.** Full length back zipper. $4.84. [$20-25] **Walking Shorts.** $3.84. [$15-20] Bottom. **Chelsea-collar Lace Outfit.** Two-piece white lace of Arnel® triacetate and nylon, fully lined in pink cotton. $8.84. [$30-35] **Embroidered Outfit.** Sleeves, skirt waistband and border embroidery, white cotton lining. Light blue. $12.84. [$30-35] Spring/Summer 1965.

Multi-stripe Cardigan. Rainbow pastels in soft knit of wool, nylon, and mohair. $10.90. [$12-15] **Cable-stitch Cardigan.** Soft knit blend of wool, nylon, and mohair. Light blue. $6.97. [$12-15] **V-neck Pullover.** Loopy stitch wool, nylon, and mohair. White. $8.97. [$12-15] **Cable-stitch V-neck Pullover.** Wool, mohair, nylon knit. Yellow. $6.97. [$12-15] **Coordinated Headband.** $.97. [NPA] Fall/Winter 1965.

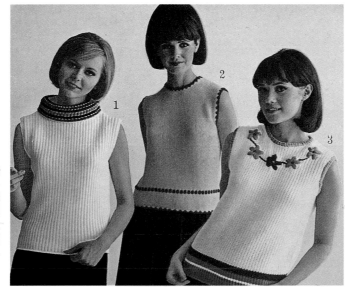

A collection of knit pullover shells in cowl-collar, embroidered trim, and floral embroidery styles. $5.94 each. [$12-15] Fall/Winter 1965.

Long sleeves boutique-print overblouses in a variety of interesting patterns. Abstract Swirl. Acetate surah screen print in olive green and white. $5.90. [$15-20] **Deep-hued Floral Print.** Acetate and nylon crepe in violet and black on royal blue. $6.70. [$15-20] **Bright Floral Print.** Acetate surah in fuchsia rose, dark red, and green on white. $3.90. [$15-20] **Soft Floral Print.** Acetate crepe in an overall gold, apple green, and orange print. $4.90. [$15-20] Fall/Winter 1965.

Pullover sweaters in wool and mohair knit. $8.94, each. [$12-15] Fall/Winter 1965.

67

Bright and colorful boutique prints are presented in these overblouses.
Diamond Plaid. Arnel® triacetate surah in coral red, royal blue, and yellow on white. $6.97. **Bowknots and Butterflies.** Acetate crepe in light blue, hot pink, and lemon yellow on white. $4.97. **Stylized Flora and Fauna.** Arnel® triacetate surah in hot pink on white. $6.97. **Floral Splash.** Acetate shantung in green, blue, and magenta on white. $6.97. Spring/Summer 1966.

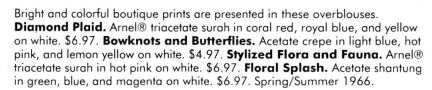

Women's Fashions

Sportswear

Turquoise and white coordinated sportswear in combed cotton knit. Tops, $2.84. [$20-25] Bottoms, $2.84-$5.84. [$15-35] Spring/Summer 1965.

Bandanna Print Cardigan-style Shirt and Kerchief Set. Reversible kerchief is solid red on other side. Combed cotton. $2.84. [$20-25] **Cabin Boy Pants.** Cotton duck with back pocket. Blue. $2.84. [$25-30] Spring/Summer 1965.

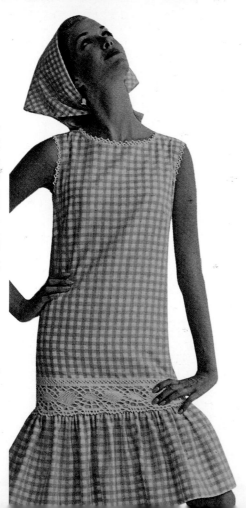

"Gingham is fresh, gingham is gay, gingham is springy." This shift with a flounced lace-trimmed hem of Dan River cotton was part of a coordinated wardrobe collection. Pink and white checks. $6.84. [$40-45] Matching triangle scarf, $.97. [$15-20] Spring/Summer 1965.

White Arnel® triacetate crepe top in a blue and green screen-print strip border design. Blue sharkskin pants, lined in cotton. Set, $9.84. [$40-45] Spring/Summer 1965.

70

Bermuda-Collar Gingham Check Shirt. Horizontal stretch Fortrel® polyester and combed cotton blend. Green and white. $3.84. [$15-20] **Capri Pants.** Stretch sailcloth with stitched creases. Green. $4.84. [$30-35] **Matching Fold-away Sling-back Slippers.** Linen-look rayon and cotton, white vinyl ankle strap with adjustable buckle. Light kelly green. $2.87 pair. [$20-25] Spring/Summer 1965.

Denim, left, and Poplin, right, styles featuring the contrasting saddle-stitching for a western look. Tops, $1.86-$2.86. [$25-30] Back-wrap skirt, $2.97. [$15-20] Culotte, $3.57. [$20-25] Shorts and pants, $1.97-$2.97. [$25-30] Jackets, $3.97-$4.97. [$25-30] Spring/Summer 1965.

Multicolor Striped Popover. Looped acrylic knit to wear with or without turtleneck sweater. Embroidered armholes and neckline. One button back closing. $5.97. [$15-20] **Fringe Poncho.** Triangular shaped of brushed nylon, black wool braid at neck, and wool fringe at hem. Wear with a slim skirt or pants, indoors or out. $9.97. [$20-25] Fall/Winter 1965.

71

Touches of the Old West make fashion news today

Popular tiny calico flower prints offer an "Old West Country Look." **Ruffled Shift.** Rayon and cotton broadcloth with white embroidery. $10.90. [$40-45] **Long-sleeved Shirt.** Cotton and rayon broadcloth with long pointed collar and crescent shaped pockets. $5.90. [$20-25] **Hip-rider Pants.** Straight-cut legs, vinyl belt. Cotton broadcloth. $6.90. [$25-30] **Blouse and Kerchief.** Button-front cardigan neckline blouse with patch pockets. "Wear triangular kerchief front-tied and buttoned onto pockets, tied in back, or as a head scarf." Cotton percale. $4.90. [$20-25] **A-line Skirt.** White cotton duck. Belt not included. $5.90. [$12-15] Spring/Summer 1966.

The hip-rider look in eyelet embroidery, and cotton and acetate hopsacking, detailed with western-style double stitching. Burgundy wine and white. **Jacket.** Simulated pearl buttons, mock pocket flaps. $13.90. [$15-20] **A-line Skirt.** Two front pockets with a watch fob pocket inside one of the pockets, fly-front zip. $8.90. [$8-10] **Ranch Style Shirt.** White eyelet embroidered cotton, simulated pearl buttons, mock pockets. $8.90. [$15-20] **Slim Hip-hugger Pants.** Pockets similar to skirt. $10.90. [$15-20] **Blouse with Kerchief.** $5.90. [$15-20] **Wool Felt Cowgirl Hat.** Rayon grosgrain band and chin strap. $3.97. [$20-25] Spring/Summer 1966.

72

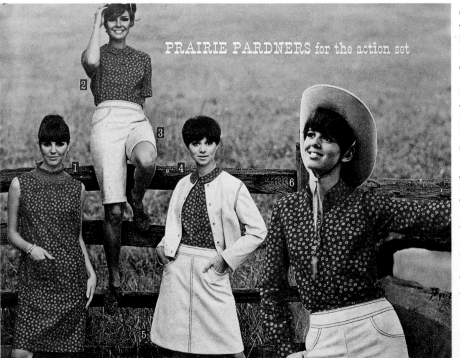

PRAIRIE PARDNERS for the action set

Calico fashions in yellow floral print on red cotton broadcloth. **Smocked Shift.** $5.90. [$20-25] **Bermuda-collar Shirt.** $3.90. [$12-15] **Cuffed Knee Length Pants.** Hip-hugging contour yoke with western-style pockets. Wheat-beige horizontal stretch denim with contrasting red stitching. $4.90. [$35-40] **Cardigan Jacket.** Gold-color metal buttons. Stretch denim in wheat-beige. $5.90. [$20-25] **A-flare Skirt.** Contour yoke. Same fabric as previous. **Button-down Collar Shirt.** $3.90. [$8-10] **Hip-rider Pants.** Wheat-beige stretch denim with contrasting red stitching. $5.90. [$20-25] **Matching Shoes.** 3-eyelet style flatties with red ribbon laces. $2.97. [$20-25] Spring/Summer 1966.

Floral-printed polyester and cotton voile coordinates in cool turquoise blue and apple green tones. **Cut-away Jacket.** Bodice lined with white voile. $7.90. [$20-25] **Skirt.** Double knife pleats. $5.90. [$15-20] **Shirt.** $7.90. [$10-12] **Walking Shorts.** Trim-fitted, with contour bandless waist, fully lined with white voile. $4.90. [$12-15] Spring/Summer 1966.

Fashions with the look of dotted Swiss in Dacron® polyester and cotton voile. All in light grass green accented with white. **Empire Shift.** Optional white cotton pique belt. $6.90. [$25-30] **Angel Top.** Pullover with white cotton pique yoke. Attached midriff-length self-liner has elasticized hem. $4.90. [$20-25] **Straight-leg Pants.** Fully lined with cotton. White pique tie-belt. $5.90. [$20-25] **Cut-away Jacket.** Bodice lined with cotton. $5.90. [$20-25] **Slim Skirt.** Fully lined in cotton. $3.90. **Buccaneer Shirt.** Pullover laced with white cotton pique. $4.90. [$20-25] **Walking Shorts.** Fully lined in cotton with white cotton pique tie-belt. $3.90. [$10-12] Spring/Summer 1966.

Tailored Shirt. Floral-printed cotton voile in hot pink, turquoise, blue, and lemon yellow. Bodice lined in cotton. $5.90. [$15-20] **Hip-rider Pants.** Pink cotton twill with contour self belt. $3.90. [$20-25] **Empire Shift.** Fully lined in cotton. $7.90. [$25-30] Spring/Summer 1966.

The bell-bottom pants with a contour, bandless waistline was a seasonal favorite. This version features front flaps in a stretch cotton and nylon denim. Navy blue. $3.90. [$30-35] Spring/Summer 1966.

Women's Fashions

Beachwear

Sears Sea Stars® swimsuits. **Sleek Crepe Sheath.** Waterproof gold color belt. Acetate/nylon/spandex blend. White. $25.80. [$40-45] **Hand-screened Floral Print Suit.** Medium green and white. $19.80. [$45-50] **Striped Sheath.** Chocolate brown on white. $18.80. [$50-55] **Gingham Check Suited Suit.** Eyelet ruffles at hemline. Candy pink and white. $9.90. [$45-50] Spring/Summer 1964.

Two-piece gingham check bikini with sheer cotton jacket with matching print border. $24.70. [$45-50] Spring/Summer 1964.

Mix and match beachwear in a blend of stretch nylon bouclé. Pieces are sold separately. **Beach Pullover.** Hot pink. $7.86. [$12-15] **Blouson Swimtop.** Pull-on style with built-in foam rubber bra. Sun yellow. $6.86. [$15-20] **Trunks.** Completely lined. Surf blue. $6.86. [$15-20] Spring/Summer 1965.

Bra Tops with built-in rubber bras available in two styles, modified bikini and high neckline. **Trunks** shown in modified bikini style and boy-leg style. In coordinated print and lime green. Each, $4.86. [$20-25] Spring/Summer 1965.

Opposite page:
Sears Sea Stars® swimsuits.
The Perfectionist.
Adjustable sheath of elasticized bengaline with shirred sides. "Flatters too thin figures - Camouflage others." Violet. $16.50. [$50-60] **"Aloha" Petal Swimcap.** Lemon yellow and 5 other colors. $4.87. [$35-40] **Woven-Stripe Skimmer.** $16.80. **Two-piece Knit Suit.** Moss green with white. $16.80. [$40-45] Spring/summer 1964.

Sears Sea Stars® swimsuits with matching tops. **Maillot with Jacket.** Tangerine floral print on white. Separate strapless bra secures to inside of suit and hooks independently. Jacket in stretch nylon yarn. $22.80. [$15-20] **Knit Two-piece with Sleeveless Top.** Brushed Orlon® acrylic in lagoon blue. $19.80. [$45-50] **Multicolor striped Cotton Suit with Cover-top.** $12.87. [$45-50] Spring/Summer 1964.

75

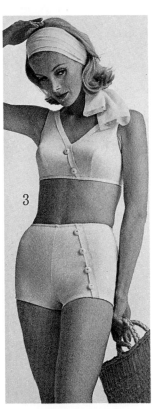

Sears Sea Stars® label swimwear. **Floral Print Tight-Leg Suit.** Nylon yarn knit in rose and white print on black. $22.70. [$30-35] **Shirred Sheath.** Acetate, cotton, and rubber in elasticized bengaline. Pre-shaped underbra. Tummy control panel. Straps adjust or can be tucked away. Blue. $16.50. [$45-50] **Tailored Short Suit.** Navy blue nylon yarn knit with white Arnel® triacetate collar, vinyl patent red belt. $19.70. [$40-45] Spring/Summer 1965.

Sears Sea Stars® label swimwear. **Knit Sheath.** Pre-formed inner bra hooks independently of suit. Elasticized waist, tummy control panel. Moss green. $18.70. [$45-50] **Skirted Two-piece Floral Suit.** Skirt has elasticized back waist with attached self-fabric underpanty. Yellow floral printed broadcloth of cotton and rayon. $13.70. [$50-55] Spring/Summer 1965.

Junior-size swimsuit in acetate, cotton, and rubber elasticized sharkskin. Bra top actually buttons in back. Pale green with blue-gray trim. $15.70. [$55-60] Spring/Summer 1965.

Sears featured "the new idea - swim tops and trunks you buy separately" to ensure perfect fit. **Striped Swim Pullover.** Nylon yarn knit. Pre-molded underbra hooks separately. Blue, green, yellow, and orange. $14.84. [$30-35] Not shown, swim bra with removable straps. $7.84. [$20-25] **Swim Trunks.** Nylon yarn knit. Green, or, not shown, blue. $6.84. [$15-20] **Three-piece Woven Plaid Swim Set.** Bra has tucked-away strap option with pre-molded inner cups, and back button closing. Hip shorts with back zipper. Poncho hood lined with blue cotton. Set, $24.70. [$50-55] Spring/Summer 1965.

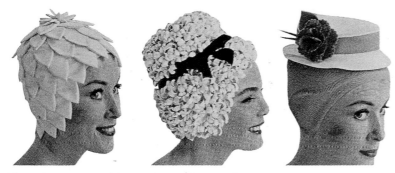

Three interesting rubber swimcap designs, featuring a Sava-wave® inner rim. **Petal Swimcap.** Lemon yellow. $4.84. [$65-70] **High Crown Bonnet.** White curled petals trimmed with black band and bow. $9.84. [$65-70] **Hat-on-a cap.** Flower trim white hat with moss green cap and trim. $4.84. [$75-80] Spring/Summer 1965.

1

Two-piece Swimsuit. Striped seersucker and blue chambray trimmed with white grommets, both of acetate and cotton. Attached cotton underpanty. Blue and white. $14.00. [$50-55] Spring/Summer 1966.

Hip-rider Swimsuit. Dot-printed cotton broadcloth with cotton lace ruffles. Loden green with white. $12.00. Spring/Summer 1966.

Blue Denim Two-piece. Western style saddle-stitched cotton and nylon denim. Shorts has mock fly-front, and attached cotton panty. $10.90. [$45-50] **Princess Sheath.** Two-way stretch knit of nylon and spandex. Coral red. $10.90. [$40-45] Spring/Summer 1965.

Women's Fashions

Maternity Wear

Formal occasion dresses for "waiting" in fine fashion. **Two-piece Lace Dress.** Cowl collar top overlaid with acetate and nylon lace. Pink. $10.84. [$25-30] **Black Jumper.** Cotton and rayon with mock pocket flaps. $6.84. [$30-35] **Black and White Printed Blouse.** Acetate surah. $3.94. [$20-25] **A-Line Dress.** Fine textured rayon with the look of linen. Rose pink. $11.84. [$25-30] **Empire Crepe Dress.** Rayon and acetate dress, elastic at front waist, rhinestone ornament. Black. $11.84. [$25-30] Spring/Summer 1964.

A variety of fashionable "wardrobe-in-waiting." The reversible coatdress pictured at left offers two dresses in one. The triangle scarf was a popular accessory during this period. The skirts of all 2 and 3-piece dresses have front cut-out and tie adjustments to accommodate the expanding belly. $4.84-$9.54. Spring/Summer 1964.

Sweeping-wide maternity dress in rayon georgette veiled over polyester crepe. Should button closing, lined in front. Aqua. $25.00. [$15-20] Spring/Summer 1964.

Pullover-style jumper shift of wide wale cotton corduroy. Deep blue. $6.94. [$15-20] Fall/Winter 1964.

Poised and pre[tty]
MATERNITY F[...]

74 SEARS CP9KM 2 DSL

Bright and flowery separates for the woman in waiting. Skirt and pants have seam to seam front stretch panel. *From left.* Schiffli embroidered popover top and skirt, Jacket with matching pedal pushers, Print duck pullover top and capri pants. Bedford cord. Coral. $2.94-$4.94. [$25-35] Spring/Summer 1964.

Sophisticated maternity fashions. Some skirts are now featuring a stretch panel in front, while others still have the front cut-out and tie adjustment. **Pink Polka-Dot Three-piece Outfit.** Linen-look rayon top and skirt, cotton broadcloth blouse. Front stretch panel on skirt. $10.84. [$30-35] **Chelsea-look Jumper and Blouse.** Jumper of textured rayon, detachable bow. Acetate crepe blouse buttons in back. Green with black trim, white blouse. $10.84. [$30-35] **Woven Check Two-piece Dress.** Arnel® triacetate and cotton, white cotton pique collar, nylon velvet bow. Aqua blue and white. Cut-out front and tie skirt. $4.84. [$25-30] **Navy Blue and White Three-piece Dress.** Cotton and rayon. Jacket with white trim. Front cut-out and tie skirt. Sleeveless white pullover blouse. $7.84. [$30-35] **Embroidery Panel Two-piece Dress.** Cotton pique, rick-rack edging. Front stretch panel skirt. White with green. $6.54. [$20-25] **Smock Trim Dress.** Dacron® polyester and cotton flock-dotted fabric with the look of dotted Swiss. Smocking trim in front and back. Copen blue with white. $6.84. [$20-25] Spring/Summer 1965.

Lace Two-piece Suit. Cotton lace permanently bonded to backing of acetate tricot. Skirt has stretch front panel. White lace on light rose tricot. $13.84. [$15-20] **Draped Neckline Dress.** Black acetate and rayon crepe bonded to acetate tricot. $10.84. [$20-25] Spring/Summer 1965.

Mandarin-collar Calico Print Shirt. Roll-up cuffs. Yellow print cotton percale. $2.94. [$20-25] **Denim Walking Shorts.** Cotton and nylon blend with horizontal stretch, and front stretch panel. Blue. $3.94. [$15-20] **Bandanna Print Pullover Top.** Cotton percale with matching triangle scarf. $3.54. [$20-25] **Denim Capri Pants.** Cotton and nylon blend with horizontal stretch, and front stretch panel. Blue. $4.94. [$15-20] Spring/Summer 1965.

Coat and Dress Ensemble. Linen-look rayon shirt-styled coat and print dress. Can be worn separately. Pink and white. $11.84. [$20-25] **Smoke Ring Collar Dress and Coat.** Nubby-textured rayon and silk. Sleeveless coat with mock pockets may be worn as a jumper. Blue, yellow, and brown on white. $8.84. [$20-25] Spring/Summer 1965.

Striped Shirt. Printed cotton broadcloth. Pink, yellow, beige, and white. $3.94. [$15-20] **Gabardine Skirt.** Dacron® polyester and combed cotton with Lycra® spandex. Front stretch panel. Rose pink. $5.94. [$15-20] **Floral Print Pullover Top.** Textured cotton. Pink, orange, and green on white. $3.94. [$25-30] **Garadine Shorts and Capri Pants.** Dacron® polyester and combed cotton with Lycra® spandex. Front stretch panel. Rose pink. Shorts, $4.94. [$20-25] Pants, $6.94. [$20-25] Spring/Summer 1965.

Lace Coat and Dress Set. Dacron® polyester and combed cotton poplin with a water-repellent, wrinkle-resistant silicone finish. Empire-style bodice of sleeveless dress with white lace and red piping, three-button front opening. Red contrasting stitching on coat. Deep copen blue. $10.90. [$20-25] **Print Coat and Dress Set.** Nubby-textured rayon and silk. Print dress to wear alone or with coat. Coat has covered buttons, mock pockets, and can be worn as a jumper. Pink and white. $9.90. [$20-25] Spring/Summer 1966.

White Lace Blouse. Acetate and nylon lace, lined with cotton except for sleeves. $3.94. [$20-25] **Black Crepe Skirt.** Acetate and rayon bonded to acetate tricot. Front stretch panel. Ankle length, $6.94. [$12-15] Street-length, $4.94. [$10-12] **Light Pink Bow Front Dress.** Rayon and Lurex® metallic knit. $12.84. [$35-40] **Pale Beige Two-piece Lace Dress.** Cotton and nylon lace lined with acetate, rayon satin bow. Skirt has stretch front panel. $13.84. [$25-30] **Brocade Suit Dress.** Two-piece cotton and acetate brocade in a floral design. Mock pockets, self-covered buttons. Skirt has stretch front panel. Copen blue and emerald green. $12.84. [$25-30] Fall/Winter 1965.

Left. **Aqua Blue Top and Skirt.** Cotton knit. Skirt lined with cotton and have a seam-to-seam front stretch panel. $3.94 each. [$15-20] **Bow-trimmed Blouse.** Combed Pima cotton crepe. Aqua blue and yellow print on white. $4.54. [$15-20] **Capri Pants.** Cotton knit, lined with cotton. Front stretch panel. Aqua blue. $3.94. [$20-25] *Right.* **Mix and Match Blue Denim and Striped Seersucker.** Skirt and pants have front stretch panel. Blue and white. $2.94-$5.84. [$30-35] Spring/Summer 1965.

A smocked dress in woven plaid cotton or blue broadcloth suitable to wear after pregnancy. Navy blue, green, red, and yellow plaid, or Light copen blue. $6.90. [$20-25] Spring/Summer 1966.

81

Women's Fashions

Loungewear and Sleepwear

Medieval look gown and embroidered blue and beige floral peignoir set. Waltz-length nylon sheer lined in nylon tricot. $29.95. [$45-50] Spring/Summer 1964.

Sleepwear sets to wear together or separately. Each sleep shift comes with a matching duster or jumper. Cotton batiste, polyester/cotton blend, or nylon tricot. $4.97-$7.97. [$20-35] Spring/Summer 1964.

Leisure shifts perfect for lounging. **Woven Plaid Shift and Skirt.** Arnel® triacetate and cotton blend. Blue. $6.77. [$20-25] **Floral Shift.** Shantung weave cotton with solid trim at hem. Pink. $5.74. [$20-25] **Free-flowing "A".** Cotton sateen floral print to wear belted or loose. $3.87. [$25-30] Spring/Summer 1964.

Colorful dusters in three styles and design patterns to "look pretty from morning coffee on." All cotton and cotton/acetate blend. $3.84-$4.84. [$15-20] Spring/Summer 1964.

Ombre Stripe pattern duster in easy-care Dacron® polyester and cotton. This style is proportioned to fit all figure types. Blue. $5.77. [$15-20] Spring/Summer 1964.

Stay-at-home leisurewear included these figure-flattering shifts. **Tapestry Modified "A" Shift.** Solid color kick pleats. Coral. $5.77. [$30-35] **Striped Shaped Skimmer.** Ticking stripe in cotton denim, white rick-rack trim. Blue. $3.77. [$20-25] **Seersucker Shift.** Embroidered flower patch pockets trimmed in rick-rack braid. Pink. $5.74. [$20-25] Spring/Summer 1964.

83

These sleepwear fashions received the Sears 4-Star Value for its laboratory-approved "powder puff soft" texture. Dacron® polyester, nylon and cotton blend. Eyelet embroidery and smocking detail. Panty-shift, baby doll, or capri pajama styles. Pink or mint green. $3.80-$4.80. [$15-25] Spring/Summer 1964.

Cotton sleepwear in three different styles, including the "new toga styling" in red cotton checks. $2.87-$3.87. [$20-35] Spring/Summer 1964.

Brush Orlon® acrylic pile robes in aqua and white peignoir style tied with satin sash, and double breasted and collarless styles with pompon buttons. Shown in Aqua, White, Teal blue, and Red. $9.77-$15.67. [$20-25] Fall/Winter 1964.

Nylon Chiffon Quilted Duster. Puffy bow of nylon tricot. Blue, green, and yellow floral pint on white. $5.74. [$15-20] **Quilted Cotton Duster.** Sectioned panel pockets across front. Zip front. Red bandanna print. $5.74. [$15-20] **Pinwale Corduroy Duster.** Embroidered border down front. Blue. $5.74. [15-20] **Wide-wale Cotton Corduroy Duster.** Available in four proportioned lengths. Red, Moss green, Pumpkin, and Bright blue. $6.99-$7.99. [$15-20] Fall/Winter 1964.

Lacy sleepwear trimmed with schiffli embroidery include this pettipant pajamas and baby doll set. Dacron® polyester and cotton blend batiste. Light blue. $2.94 each set. [$20-25] Spring/Summer 1965.

YOUNG JUNIOR SLEEPWEAR

Soft lustrous batiste in an easy-care blend of Dacron® polyester-and-cotton. Wrinkle-resistant and shrinkage controlled. Machine washable . . medium; machine dry or drip dry.

A variety of fashions for lounging around. Shown are cotton seersucker A-line shifts, rayon and cotton smocks, and cotton full length granny shifts. $3.90-$6.80. [$12-25] Spring/Summer 1966.

Sleepwear ranging from the provocative to the granny look. **Ripply Ruffles.** Aqua or Red nylon tricot. $4.97. [$10-12] **Caged Bikini.** Nylon tricot animal print bikini, nylon fishnet "cage" bound with matching print. $3.90. [$40-45] **Romper.** Nylon tricot one-piece. Maize or Blue. $3.90. [$20-25] **Tennis Dress.** Aqua or Apricot nylon tricot, trimmed in white. $3.90. [$20-25] **Candy-stripe Romper.** Nylon tricot one-piece with a two-piece look. Peppermint pink or Mint julep. $5.80. [$30-35] **Panty Shift.** Cotton with lace trim bib and sleeves. Includes panty. $3.90. [$15-20] **Polka-dot.** Includes shirt, panty, and curler cap. Pink. $3.90. [$15-20] **Matching Boots.** $2.22. [$12-15] **Bikini Ensemble.** Nylon tricot with lacy ruffles. Bikini and matching cover-up coat. Red, Black, or Aqua. $5.80. [$15-20] Spring/Summer 1966.

Women's Fashions

Fashion Accessories

Designer hats by Mr. John, each packed in its own box. Notice the puffed crown look. *Left.* Mr. John Jrs. Collection for the well-dressed look. **Fascinating Fedora.** Rayon/cotton/acetate. Natural. $20.97. [$45-50] **Packable Fedora.** Brim can be positioned up, down, or to the side. Woven Swiss Rayon Felt. Navy with white. $17.97. [$45-50] **Bloused Pure Silk Roller.** Pink dots on white. $24.97. [$50-55] **Print at the Top.** Arnel® triacetate surah print with straw braid. Black and white. $24.97. [$50-55] *Right.* Mr. John Classics. **Ripple Brim.** Shantung of rayon and silk. Pink. $12.97. [$45-50] **Extravagant Floral Print.** Yellow, green, and white acetate. $12.97. [$50-55] **Lattice Picture Hat.** White textured straw with black. $12.97. [$75-85] Spring/Summer 1964.

Imported pre-styled wigs means there's never a bad hair day. "Career girl or busy mother, vacationer or sportswoman - for every woman on the go, a wig is an asset." Wig block, storage box, and instructions included. Choose between the economy human hair made in Japan, $69.50 [NPA] or the higher priced 100% European Hair made in West Germany, $119.95. [NPA] The higher priced wig has softer, finer textured hair and comes in two sizes to assure correct fit. Spring/Summer 1964.

Designer hats by Lilly Daché for the well-dressed woman. Each hat packed in its own designer hat box. Left. **Eye-Level Cloche.** Viscose rayon, shantung-textured. Turquoise with white trim. $19.97. [$50-55] **Forward-Tilt Straw Cap.** Red with navy. $18.97. [$65-75] **Alpaca Cloche.** Pink with white. $18.97. [$45-50] Center. **Sportive Swagger.** Straw braid with rayon grosgrain ribbon. Silver gray with medium gray and red. $21.97. [$85-90] Right. **Winged Toque.** White straw braid with black patent trim. $18.97. [$45-50] **Cuffed Rayon Bouclé.** Beige with coffee. $19.97. [$45-50] **Brimmed Breton.** Ribbon tabbed straw braid. White with navy. $18.97. [$45-50] Spring/Summer 1964.

A collection of lower priced Sears hats featuring flowers and netting. $1.97-$5.97. Spring/Summer 1964.

Sally Victor designer hats are packed in its own designer box. *Left.* Sally Victor Headlines Collection in straw. **Bolivian-style Sailor.** Brown with tri-color trim. $12.97. **Straw Helmet.** Black with white and red. $12.97. **Gaucho Sailor.** Black. $12.97. **Pixie Pillbox.** Navy, light blue, and white. $12.97. *Right.* Sally Victor Headlines Collection. **Twist Turban.** Beige, black, and white. $14.97. Sally V Collection. **High Round Straw Dome.** White with navy and red. $18.97. **Sweeping Fur Felt.** White with black stitching. $14.97. Spring/Summer 1964.

Emme Boutiques designed by Emme. All hats come packed in its own designer hat box. *Left.* **High-hat Straw Roller.** Black with white trim. $18.97. [$25-30] **Shadow-brimmed Stovepipe.** White straw braid with beige plastic patent band. $20.97. [$40-45] **Dramatic Turban.** Draped woven nylon tubing, alpaca cuff and bow. White. $20.97. [$20-25] *Center.* **Cartwheel Hat.** Black straw braid, alpaca edging. $21.97. [$45-50] *Right.* **Flowered Beehive**. Pink ribbed imported straw braid, pink alpaca brim, red rose. $19.97. [$20-25] **High High Crown Casual Cloche.** Yellow ribbed imported Swiss straw. $19.97. [$20-25] **Ribbed Straw Roller.** Blue. $19.97. [$30-35] Spring/Summer 1964.

Designer label *Dachettes by Lilly Daché* comes with its own hat box. **Velour Fur Felt Boater.** Coffee brown with beige rayon satin trim. $24.97. [$35-40] **Rayon Velvet Scoop.** Black. $24.97. [$55-60] **Dressy Rayon Panné.** White with black dots and brim. $22.97. [$50-55] Fall/Winter 1964.

Spring hats in big-brim, profile, cloche, pillbox, roller, swagger, and turban styles. $3.97-$6.97. [$15-35] Spring/Summer 1964.

Pace-Setters by Mr. John. All hats arrive in its own designer box. *Left.* **Tailored Profile.** Velour felt fur. Blue and green. $14.97. [$45-50] **Draped Turban.** Fluffy mohair and bands of rayon velvet. Cranberry red. $24.97. [$40-45] *Center.* **Wool Felt Bobby.** London-inspired. Black with white trim. $8.97. [$55-65] **Slouch.** Velour fur felt. Green with white trim. $12.97. [$45-50] **Big Bow Profile.** Beaver fur felt. Medium gray. $24.97. [$55-65] Fall/Winter 1964.

PACE-SETTERS BY *Mr. John*

Sally Victor Headliners label designer hats in velour fur felt. Each packed in its own box. **New Padre Shape.** Accented with feather. Gray. $16.97. **Peruvian Style Sailor.** Wide braid and ribbon band. Navy with red trim. $14.97. **The Derby Suiter.** Banded with plastic patent. Black with red band. $12.97. **Sophisticated Profile.** Natural pheasant feather. Coffee brown. $15.97. Fall/Winter 1964.

Emme Boutique designer hats, in its own box. **Big Brim.** Velour fur felt with snakeskin band. Moss green with red/orange band. $21.97. [$45-50] **Rich Textured Sailor.** Beaver fur felt, brim edged with rayon satin. Black. $20.97. [$40-45] **Off-the-face Beret.** Corded rayon trimmed with capeskin cuff and back bow. Black with green trim. $18.97. [$40-45] **Sugar Scoop Suiter.** Velour fur felt with snakeskin band. Turquoise blue with coffee brown band. $21.97. [$55-60] Fall/Winter 1964.

89

Styled by *Amy New York* designer hats, each with its own box. *Top.* **Feather Trim Large Cloche.** Velour fur felt. Emerald green with sapphire blue trim. $17.97. [$30-35] **Profile.** Velour fur felt in black. $16.97. [$30-35] **Bowler.** Beaver fur felt in Beauty red. $12.97.[$30-35] Fall/Winter 1964.

Sears generic label hats accented with fur and feathers. Natural furs priced to include a 10% Federal Excise Tax. *Left to right, top to bottom.* **Marabou Feather Tamourine.** White. $5.97. **Tall Story Feathers.** Red. $10.97. [$35-40] **Feather Pouf Flatterer.** Gold color with natural pheasant. $6.97. [$40-45] **Natural Mink Tail Pillbox.** Includes anchor combs. Natural ranch mink. $18.97. [NPA] **Natural Mink Tail Pin (on lapel).** Natural ranch mink with gold color center. $1.97. [NPA] **Natural Mink Fur Pillbox.** Fine rayon matte jersey top and cuff. Natural pastel mink. $15.97. [NPA] **Natural Mink Tail Halo.** Four mink tails on rayon velvet with veil. Natural pastel mink. $5.97. [NPA] **Fur "Wig" Hat.** Natural sable tails on wool knit base. $12.97. [NPA] **Glamour Brocade Turban.** Rayon and Mylar® metallic with mock jewels. Blue, beige, and brown. $6.97. [$20-25] **Sequined Cloche.** Iridescent glow hat with rayon grosgrain band. Turquoise blue. $8.97. **Brocaded Dome Pillbox.** Black with gold and silver color metallic brocade. $4.97. [$20-25] Fall/Winter 1964.

Styled by *Betmar* hats comes packed in its own box. **Soft Brim Cloche.** Rayon velvet with stitched silk jersey brim facing. Black with beige trim. $14.97. [$40-45] **Flip-Brim Cavalier.** Beaver fur felt trimmed with ribbon. Teal blue. $14.97. [$55-65] **Swagger Fedora.** Fur felt with stitched indented crown. Beige. $12.97. [$30-35] **Ripple-brimmed Cloche.** Fur felt stitched to retain shape. Coffee brown with black trim. $12.97. [$40-45] **Sailor-brimmed Bowler.** Fur felt in beige with coffee trim. $10.97. [$45-50] Fall/Winter 1964.

An assortment of women's fashion headwear priced low. $1.97-$7.97. [$20-35] Fall/Winter 1964.

Sears fashion headwear individually sized in small, medium, and large for that perfect fit. $3.97-$8.97. [$15-35] Spring/Summer 1965.

Winter hats that sport both warmth and beauty. Priced low from $1.97-$8.97. [$20-55] Fall/Winter 1964.

A collection of Sears hats sized in small, medium, or large. The four styles at right is one size fits all. The interesting white combination veil, halo ring, and bow can be worn separately to create new looks. Hats, $2.97-$4.97. [$15-35] Combination Hat and Handbag Set, $12.97. [$25-40] Spring/Summer 1965.

These Hats fit all

9 Hand-crocheted snood of rayon ribbon. Flat front bow.

Fashion headwear shown in numerical order. **Wool Felt Cloche.** Coffee brown with coffee, brown, and white ribbon trim. **Feather Top-knot.** Lack rayon velvet ring, veil, and iridescent coque feather. **Wide Brim Velvet.** Multicolor with black. **Cranberry Red Turban.** Rayon velvet toque with rayon taffeta sides. **French Profile.** Gold wool felt. **Wool Felt Casual.** Cranberry red with rayon grosgrain ribbon trim. **Packable Snood.** Rayon velvet with wool jersey band. Light beige. **Rayon Velvet Pillbox.** Bright red with veil. **Rayon Velvet Whimsey.** Black, with dotted veil. **Marabou Feather Toque.** White with turquoise blue. **Coque Feather Half-hat.** Silver color, with rhinestone sparkle trim. **Feather Pixie.** Sapphire blue. **Brocade Roller.** Gray and blue rayon brocade crown with gold-color metallic, black rayon velvet brim. **Chenille Roller.** Wool body covered with rayon and cotton chenille. Royal blue and emerald green. **Colorful Sailor.** Multicolor rayon velvet with stitched black brim. **Chenille Beret.** Wool body covered with rayon and cotton chenille. Light beige and coffee brown. All hats are $1.97-$8.97. [$15-40] Fall/Winter 1965.

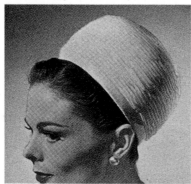

Pleated sheer nylon pillbox with rayon taffeta cuff. Pink. $4.97. [$20-25] Spring/Summer 1966.

School Year Pin shows the year in gold-color numerals on black plastic, with bright Scotch plaid wool tabs attached. Approximately 3 ½" long. Price includes a 10% Federal Excise Tax. Available for school years '65, '66, '67, and '68. $.73. [$15-25] Fall/Winter 1964.

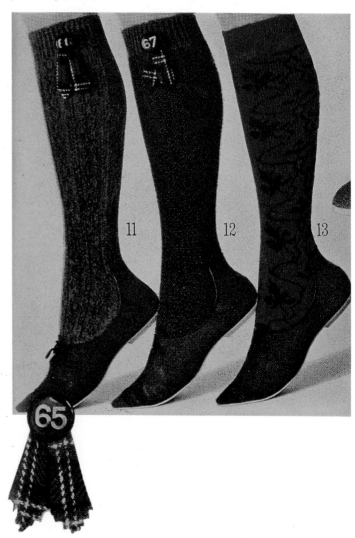

92

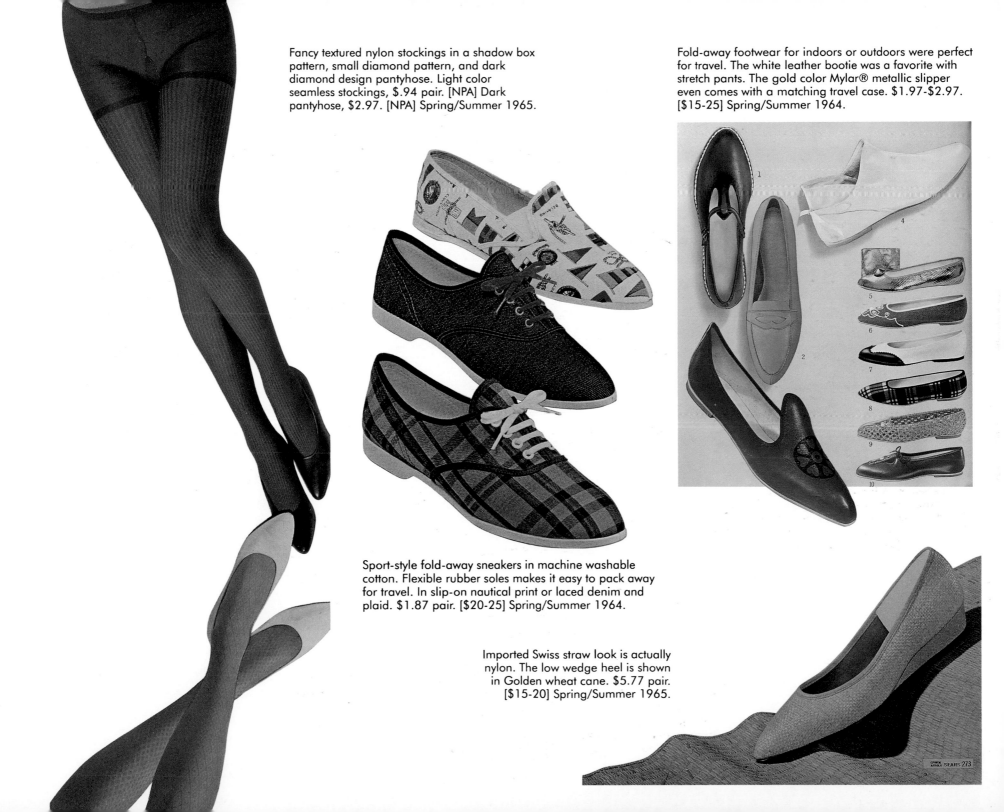

Fancy textured nylon stockings in a shadow box pattern, small diamond pattern, and dark diamond design pantyhose. Light color seamless stockings, $.94 pair. [NPA] Dark pantyhose, $2.97. [NPA] Spring/Summer 1965.

Fold-away footwear for indoors or outdoors were perfect for travel. The white leather bootie was a favorite with stretch pants. The gold color Mylar® metallic slipper even comes with a matching travel case. $1.97-$2.97. [$15-25] Spring/Summer 1964.

Sport-style fold-away sneakers in machine washable cotton. Flexible rubber soles makes it easy to pack away for travel. In slip-on nautical print or laced denim and plaid. $1.87 pair. [$20-25] Spring/Summer 1964.

Imported Swiss straw look is actually nylon. The low wedge heel is shown in Golden wheat cane. $5.77 pair. [$15-20] Spring/Summer 1965.

Genuine reptile handbags and shoes. Prices of handbags include a 10% Federal Excise Tax. **Alligator-Lizard.** Sport rust. Shoes, $14.97. [$45-50] Bag, $20.97. [$65-95] **Patched Lizard.** Brown ombre, brown heel. Shoes, $7.74. [$50-55] Bag, $9.97. [$75-110] **Alligator.** Sport Rust. Shoes, $29.95. [$50-55] Bag, $49.97. [$95-150] Fall/Winter 1964.

Thong sandals with flexible heel and sole that fold away for traveling. Antique brown leather, $2.97 pair. [$20-25] White vinyl, $2.87 pair. [$20-25] Spring/Summer 1965.

Packable, fold-away footwear in cotton denim, suede leather, and soft glove leather. Each pair comes with a clear snap-front pouch. $1.87-$2.87. [$10-12] Fall/Winter 1964.

94

More fold away shoes. **Bandanna-print Flatties.** Red print. $2.97. [$20-25] **Tie-back Boots.** Leather uppers. Cork tan. $2.87. [$15-20] **Polka-dot Boots.** Cotton denim upper, white vinyl trim. Navy with white dots. $2.47. [$25-30] **Classic Stretch Boots.** Stretch denim uppers of acetate, cotton, and rubber. Light blue. $2.97. [$15-20] Spring/Summer 1965.

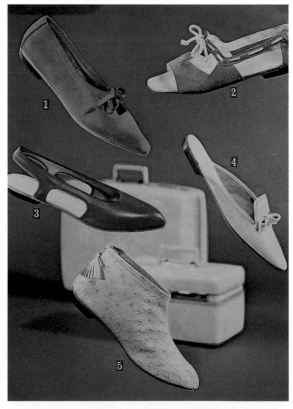

Each of these leather upper slip-on footwear features the popular flexible heel and sole, perfect for traveling light. All styles packed in its own plastic pouch. $2.97-$3.67. [$15-20] Spring/Summer 1966.

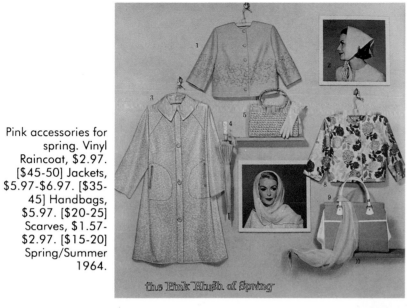

Pink accessories for spring. Vinyl Raincoat, $2.97. [$45-50] Jackets, $5.97-$6.97. [$35-45] Handbags, $5.97. [$20-25] Scarves, $1.57-$2.97. [$15-20] Spring/Summer 1964.

Samples of pettipants, "the perfect underfashion for your slimmest dresses." In nylon tricot. $1.87-$2.84. [$25-30] Fall/Winter 1964.

Fresh bloom of yellow accessories. Clear vinyl polka dot raincoat, $3.97. [$50-55] Matching umbrella, $2.97. [$25-30] Cardigan, $5.97. [$12-15] Striped Jacket, $2.97. [$15-20] Handbags, $5.47-$8.97. [$25-30] Fringed Stole, $5.97. [$12-15] Scarf, $1.97. [$15-20] Spring/Summer 1964.

A variety of jungle motif fashion accessories. **Raincoat and Kerchief.** Water-repellent acetate, waterproof vinyl lining. $5.47. [$85-110] **Jacket.** Printed rayon pile on cotton back, convertible collar, lined. $14.97. [$50-65] **Contour Belt.** Printed rayon pile on cotton back, leather lined. Gold color metal buckle. $1.97. [$35-40] **Suspenders.** Printed rayon pile on cotton back, adjustable metal clips. $1.97. [$35-45] **Tabbed Pouch.** Printed rayon pile on cotton back, rayon faille lining. Price includes a 10% Federal Excise Tax. $8.97. [NPA] **Umbrella.** Printed acetate cover, vinyl-covered handle and strap. Made in Italy. $3.97. [$35-40] **Glove and Clutch Set.** Rayon pile trim stretch wool jersey gloves. Bag is rayon lined. Price includes a 30-cent Federal Excise Tax. $5.49. [$45-55] **Beret and Boa Set.** Printed rayon pile, completely lined. Boa has double snap closing. $3.97. [$40-45] **Silk Chiffon Scarf.** Made in Japan. "Can be worn several ways - as an ascot or head scarf." $1.97. [$35-40] Fall/Winter 1964.

95

Women's Fashions

Suits

Elegant two-piece suit, richly textured tussah of hand-spun silk, fully lined. Hand loomed in India. Natural with brown top-stitching. $49.50. [$20-25] Spring/Summer 1964.

Three-piece sleeveless floral suitdress of hand-screened grainy-textured rayon, fully lined. Jacket has mock pockets. $22.00. [$20-25] Spring/Summer 1964.

Four-piece ensemble featuring a reversible wool coat with camel tan flannel on one side and white boucle on other, wool flannel sleeveless suit top and matching skirt, and polyester crepe blouse. Camel tan and pale cream. $57.50. [$20-25] Spring/Summer 1964.

Sophisticated 2-piece suit in slubbed silk with very fine twill weave. Lined jacket with shirtcuff sleeves. Sear's *Claude Riviere Collection made in the USA*. Rose pink. $29.50. [$50-60] Spring/Summer 1964.

Madrid-inspired linen-look three-piece ensemble in tightly woven rayon/polyester. Lined skirt and jacket, back of overblouse curves to V-closing with button closing. "Very 1964." Lemon yellow. $38.00. [$15-20] Spring/Summer 1964.

Tweed-look worsted wool double-knit 3-piece suit imported from England. Jacket has mock pockets, and stitching that follows curve of front. Gold color wool blouse with front drawstring at neck. Gold/beige tones. $39.50. [$35-40] Spring/Summer 1964.

97

Classic 3-piece suit imported from England. French Blue Knit with Embroidered Blouse. Smooth flat worsted wool double knit. $39.50. [$25-30] Spring/Summer 1964.

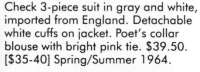

Check 3-piece suit in gray and white, imported from England. Detachable white cuffs on jacket. Poet's collar blouse with bright pink tie. $39.50. [$35-40] Spring/Summer 1964.

Classic, with day-into-evening possibilities. A smooth flat knit. Embroidered blouse in contrasting color that matches collar and facing of jacket. Hip-lined skirt zips at side; blouse zips in back. Dry clean. Dachette hat sold on page 152. Misses' sizes 10, 12, 14, 16, 18. Shorter Women's sizes 12½, 14½, 16½, 18½, 20½, 22½. N31 K 8317F—French blue with white N31 K 8318F—Black with bone beige State size. Shpg. wt. 2 lbs. 4 oz...$39.50

Floral embroidery trims the white jacket on this 3-piece double knit cotton suit. Blouse with cap sleeves buttons in back. Styled in five sizes for that perfect fit: Juniors, Misses, Shorter Women, Petite Juniors, and Tall Misses. Teal blue and white. $13.50. [$35-40] Spring/Summer 1964.

Two styles of suitdresses in Dacron® polyester and combed cotton poplin, proportionately sized to fit Petite Juniors, Juniors, Misses, or Tall Misses. These outfits are completely machine washable. Each, $7.84. [$15-20] Spring/Summer 1964.

Pastel-toned suits in three different styles. **Subtly Fitted Suit.** Worsted wool and silk. Jacket fully lined, skirt lined to below hips. Light taupe brown. $33.50. [$55-60] **Longer Jacket Suit.** Diagonal weave wool and nylon. Jacket fully lined, skirt lined to below hips. Pink. $29.90 [$65-70] **Mink Collar Suit.** Loop-textured wool bouclé. Lined jacket, seat-lined skirt. Medium blue with light brown natural mink. $39.90. [$60-65] Spring/Summer 1964.

Two-tone Stripe. Cotton double-knit in two-tone Aqua green stripe. $15.70. [$45-50] **Stripe Trim Three-piece Suit.** Worsted wool double-knit in Geranium rose and white. $29.70. [$20-25] Spring/Summer 1964.

The subtly textured spring suits in imported French acetate yarns were tested in the Sears laboratory for their ability to resist wrinkling and shape retaining features. **Casual Two-piece.** Jacket has convertible collar. Royal blue. $25.70. [$50-55] **Open Jacket Three-piece Suit.** Fine loop-scallop detailing. Yellow. $29.70. [$30-35] Spring/Summer 1964.

Sears International Collection features these suits imported from Italy. **Beige Double Knit.** Blouse and wool jacket trim in gold-color knit Lurex® metallic. $37.50. [$15-20] **Two-tone Blue Double Knit Wool.** Back zip pullover blouse and jacket accented with two tone blue. $44.50. [$15-20] **Mohair Jacket Suit.** Jacket of mohair, wool, and nylon knit. Pullover top and skirt of double-knit wool. Lilac, pink, aqua and white jacket, lilac top and skirt. $36.50. [$15-20] Fall/Winter 1964.

Couture fashions by Claude Riviere, designed in Paris, made in the USA. **Bouclé Suit.** Nub-textured wool/nylon blend, fully lined. Celery green. $59.50. [$45-50] **Wool Melton Cloth Coat.** Acetate satin lined. Putty green. $79.50. [$30-35] Fall/Winter 1964.

Three-piece Tweed Suit. Double-knit worsted wool imported from England. Dark and medium green tweed. $44.50. [$60-65] **Bright Pink Three-piece Suit.** Double-knit worsted wool in color-flecked textured plaid. Imported from England. $44.50. [$30-35] Fall/Winter 1964.

The versatile four-part suit in crisp linen-look rayon features a skirt and jacket in solid color medium blue, and a second skirt and shell in a blue and gold print on white. Set, $24.90. [$50-55] Spring/Summer 1964.

Elegant suits with a touch of real fur. **Wool Tweed Walking Suit.** Gray-dyed opossum collar. Fully lined and interlined. Red with black tweed. $54.50. [$35-40] **Wool Bouclé With Natural Mink Collar.** Dark brown nubby wool bouclé with dark brown ranch mink. $59.50. [$40-45] Fall/Winter 1964.

Three-piece Chelsea-collared suit in double-knit, multicolor-flecked tweed of French acetate fibers, nylon and wool. Trim on collar and mock pockets matches shell. White and gold tweed with gold. $29.50. [$20-25] Spring/Summer 1965.

Three-piece suit features a sleeveless overblouse and skirt of double-knit cotton, and a fully lined jacket of cotton, mohair, and nylon. Sized to fit Junior, Miss, Shorter Women, and Tall Miss. Bright rose and white. $15.75. [$45-50] Fall/Winter 1964.

Lightweight suits of woven Dacron® polyester and cotton that are never out of season. **Crisp Whipcord Jacket and Skirt.** Sleeveless batiste print blouse in blue, green, and aqua on white. Jacket features matching print lining. Blue. $17.90. [$20-25] **Round Collar Ribbed Ottoman Suit.** Jacket lined. White. $14.90. [$30-35] **Poplin Cutaway Jacket and Skirt.** Twin rows of white top-stitching. Jacket lined, with mock pockets. Light red. $12.90. [$25-30] **Blue and White Seersucker Suit.** White cotton pique blouse. Chelsea-collared jacket is lined. $17.90. [$20-25] Spring/Summer 1965.

Sears *Mary Lewis*® *Classics* in luxurious fabrics. Special features include hand-bound buttonholes with extra button included, and skirts fully lined in acetate taffeta. $29.90 each. [$20-35] **Pastel Tweed.** Wool and nylon with braid trim. Jacket lined with acetate crepe that matches blouse. Pink and white. **Loopy Bouclé Suit.** Wool and nylon. Jacket is lined and trimmed with rayon and silk shantung that matches sleeveless shell. Yellow. **Nubby Textured Wool Bouclé.** Standaway collar jacket with dome buttons. Jacket and skirt lined in acetate crepe. White. *Inset.* **Natural Mink Boa.** About 37 inches long. Rayon velvet backing. Price includes a 10% Federal Excise Tax. Pearl (light beige). $39.90. [NPA] Spring/Summer 1965.

Elegant Turquoise Blue Three-piece Suit Dress. Rayon and silk shantung, with blouse in soft rayon chiffon. Fully lined with acetate taffeta except for jacket sleeves. $19.90. [$35-40] Spring/Summer 1966.

Textured double-knit cotton, with narrow braid trim on jacket. Note the elbow-length jacket sleeves. $10.00. [$20-25] **Black-dyed Red Fox Pillbox.** Rayon satin top. One size fits all. $14.97. [$30-35] Fall/Winter 1965.

Women's Fashions

Outerwear

Hand-knit seven-eighths topcoat of mohair and wool, lined with silk. Imported from Hong Kong. White. $39.70. [$25-30] Spring/Summer 1964.

Cotton Poplin Shirt Jacket. Metal buttons, flapped chest pockets. Red with white contrasting stitching. $4.87. [$20-25] **Cotton Poplin Zip Front Jacket.** Drawstring hood and bottom. Blue with contrasting white trim. $4.87. [$15-20] Spring/Summer 1964.

Cotton Barathea Jacket. Black contrast stitching on light green, lined in bright print. $7.87. [$20-25] **Hooded Pullover.** Cotton duck with drawstring hood and hem. Royal blue with contrasting white stitching. $3.87. [$15-20] Spring/Summer 1964.

Black and white plaid "sportive" coat in soft fine wool, completely lined. $23.50. [$30-35] Spring/Summer 1964.

Glamorous jacket with a moiré pattern, turn-back cuffs, hook closing. Rayon pile on cotton. White. $18.50. [$40-45] Spring/Summer 1964.

Mary Lewis Classics were available only at Sears. The coats and suits featured fit and tailoring, scaled to fit the shorter woman, the average miss, and tall miss sizes. The styles shown are in medium blue nubby-textured wool bouclé with acetate/rayon crepe linings. **Convertible Collar Coat.** $24.90. [$35-40] **Open Jacket Suit.** Includes white rayon blouse. $29.90. [$20-25] **Notched Collar Coat.** $24.90. [$45-50] Spring/Summer 1964.

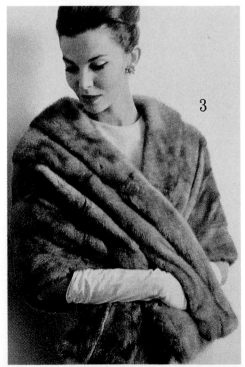

Cape-style natural mink stole with silk and rayon lining. Shown in pastel (medium brown). Also available in ranch (dark brown). Price of fur includes a 10% Federal Excise Tax. $199.00. [NPA] Spring/Summer 1964.

The "newest fashion looks" in spring coats. **Tailored Coat.** Diagonal weave wool, fully lined. Light pink. $23.50. [$30-35] **A-Shape Cardigan Coat.** Detachable cowl-collar. Nub-textured wool, fully lined. Yellow. $19.90. [$35-40] Spring/Summer 1964.

These raincoats came with their own fashion coordinated umbrellas. Shown in solid color with print lining, and madras plaid with solid gold-color lining. $12.90-$22.90. [$20-35] Spring/Summer 1964.

These glamorous water-repellent summer coats are foam-laminated for shaping and lightweight warmth. **Puckered Nylon Flare.** Rayon velvet collar and buttons. Black. $18.90. [$50-55] **A-Shape Brocade.** Cotton and acetate. Light beige. $17.50. [$20-25] **Classic Coat.** Matching color polka dot lining. Light aqua blue. $15.50. [$20-25] Spring/Summer 1964.

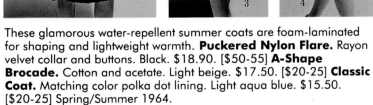

This Chesterfield plaid coat was a popular style in the *Sears Weather-Guard Classic Collection.* The fully water-repellent coat was available with or without a zip liner of Orlon® acrylic pile on cotton back. Black and olive shadow plaid. With liner, $17.94. Without liner, $9.94. Spring/Summer 1964.

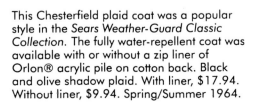

Three part cape outfit that can be worn separately. Cape in black and white herringbone tweed of wool and nylon, red lining of acrylic pile, black leather trim. Matching tweed skirt is fully lined. Black bulky knit sweater is Orlon® acrylic with mock turtleneck. $49.90. [$40-45] Cape only, $39.90. [$30-35] Fall/Winter 1964.

Wool Knit Two-piece Outfit. Blue pullover shift with contrasting beige stripes, slip-through self-tie. Beige blouse ringed with blue at collar and cuffs. Imported from Italy. $29.50. [$15-20] **Corduroy Cape Outfit.** Camel tan cape and skirt of triple-ribbed wide wale cotton corduroy. Cape has toggle closing, and is fully lined with blue and brown paisley printed combed cotton twill that matches blouse. Cape and skirt set, $18.90. [$40-45] Blouse, $4.90. [$8-12] Matching hat, $4.97. [$20-25] **Iridescent Blue Raincoat.** Rayon and acetate twill, fully lined. Fall/Winter 1964.

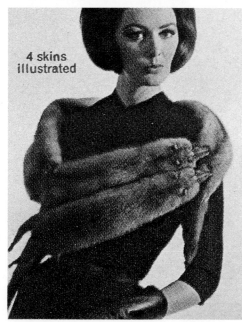

4 skins illustrated

Natural brown mink 2, 3, 4, and 5-skin scarves. Each skin is about 25 inches long from tip of nose to tip of tail. A 10% Federal Excise Tax is included in the price. 2-skin, $65.00. [NPA] 3-skin, $95.00. [NPA] 4-skin, $123.00. [NPA] 5-skin, $149.00. [NPA] Fall/Winter 1964.

Three-piece Jacket Set. Orlon® acrylic pile on cotton back jacket, lined with quilted rayon twill. Hat and mittens complete the set. White. $14.90. [$30-35] **Cotton Corduroy Hooded Jacket.** Body of jacket lined in cotton-backed acrylic pile, sleeves lined in quilted acetate. Gold with white fur trim. $14.90. [$40-45] Fall/Winter 1964.

Below, top left. **Stretch Style Nylon Taffeta Ski Jacket.** Body quilted to Dacron® polyester with cotton backing, nylon lined. Knit trim at collar and cuffs. Hood folds into back neckline. Aqua, yellow, and brown print. $19.90. [$20-25] *Right.* **Reversible Fur Trimmed Hooded Ski Jacket.** Blue and green cotton poplin print reverses to yellow nylon taffeta quilted to Dacron® polyester. Hood trimmed in black tip-dyed bleached white lamb (fur origin: Turkey). One pocket on nylon side. $16.90. [$30-35] **Reversible Nylon Ski Jacket.** Nylon taffeta quilted to Dacron® polyester reverses to smooth nylon taffeta. Knit trim. Drawstring hood rolls into collar. White and aqua. $13.90. [$20-25] Fall/Winter 1964.

Cotton Suede Cloth Jacket. Lined and trimmed with butter-color cotton backed acrylic pile. Sleeves lined in quilted acetate. Green. $13.90. [$15-20] **Suede Jacket.** Matching color zip-out liner of acrylic pile on cotton back. Contrasting black capeskin trim. Rayon lined. Beige and brown. $34.90. [$50-65] Available without zip-out liner for $5.00 less. Fall/Winter 1964.

Cherry red seven-eighths topcoat of mohair and wool is hand-knit and fully lined with silk. Imported from Hong Kong. $38.50. [$45-50] Fall/Winter 1964.

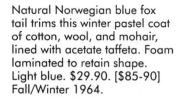

Natural Norwegian blue fox tail trims this winter pastel coat of cotton, wool, and mohair, lined with acetate taffeta. Foam laminated to retain shape. Light blue. $29.90. [$85-90] Fall/Winter 1964.

Cardigan-style cable-knit topcoat in seven-eighths length, with ¾ length sleeves. Orlon® acrylic knit with gold-color metal buttons. Royal blue. $12.90. [$20-25] Fall/Winter 1964.

Winter pastels in three styles. **The Furred Coat.** Bleached white fur collar (fur origin: Canada) on nub-textured wool/mohair/nylon bouclé. Pale green. $39.90. [$75-80] **Gentle-Look Coat.** Soft wool fleece with napped surface, convertible collar. Pale pink. $34.90. [$40-45] **Little Overcoat.** Cotton/wool/mohair laminated to foam. Pale blue. $19.90. [$45-50] Fall/Winter 1964.

Sears classic imported leather outerwear in soft luster suede that sponges clean. **Short Coat.** Wood buttons, fully lined, includes acrylic pile zip-liner. Medium brown suede and pile. $39.90. [$35-40] **Full Length Coat.** With modacrylic pile cotton backed zip-liner striped to look like fur. Dark brown suede, brown and beige pile. $69.50. [$35-40] **Full-skin Natural Mink Collar Coat.** Acetate satin lining. Two way collar closes high or lies flat. Beige. $99.50. [$55-60] Fall/Winter 1964.

The contemporary camel hair coat is made of imported fibers, lined in rayon crepe-back satin. Wooden buttons, convertible notched collar. Worn with or without belt. $55.50. [$95-100] Fall/Winter 1964.

A gorgeous fake in subtly shaded, cotton-backed modacrylic pile with added fur fibers for softness. Hood collar, full length, lined coat. Furrier clean. Medium beige. $59.90. [$55-60] Fall/Winter 1964.

Nylon taffeta jackets in pullover solid and zip-front printed styles. Each jacket is lightweight and water-repellent. Solid, $4.84. [$20-25] Print, $5.84. [$20-25] Spring/Summer 1965.

Elegant fur coats from Sears could be purchased on monthly installments with no money down. The natural mink coats shown here were densely furred and the pelts artfully matched to create the ultimate look in fashionable outerwear. Each coat is lined in silk and rayon. Pastel-medium brown. Coat, $1079.00. [NPA] Jacket, $599.00. [NPA] Fall/Winter 1964.

Natural mink stoles in cape and shawl collared styles. Each lined in silk and rayon. Pastel-medium brown. Available in the more expensive let-out skins model, or the lower priced split skins design. *Left*. **Cape Stole.** Let-out skins, $499.00. [$15-20] Split skins, $299.00. [$15-20] *Right*. **Shawl-collared Stole.** Let-out skins, $599.00. [NPA] Split skins, $399.00. [NPA] Fall/Winter 1964.

These coat styles offered sizes scaled to fit Junior, Miss, the Shorter Woman, and Tall Miss. **Rich Wool Duvetyn Tailored Coat.** Fully lined. Pink. $26.90-$28.90. [$45-50] **Wool and Cashmere Clutch Coat.** Lined. Medium blue. $19.90-$21.90. [$65-85] **Pure Cashmere Luxury Coat.** Fully lined. Beige. $37.90-$39.90. [$85-110] Spring/Summer 1965.

Houndstooth Check Flared Coat. Front is flared, back is straight and slim. Black and white check wool, fully lined. $22.90. [$30-35] **Fitted Flared Coat.** Wool and nylon diagonal weave, belted panel back, fully lined. Light pink. $22.90. [$35-40] **Double-breasted Tweed.** Wool and nylon tweed swings to a flare hemline, fully lined. Light green tweed. $22.90. [$25-30] **Impeccable Chic Coat.** "One of the smartest way to look this season." Textured basket weave wool trimmed with color matched binding. Fully lined. White. $24.90. [$25-30] Spring/Summer 1965.

Opposite page:
Spring coats with the "great new young look." **Double-breasted Tailored Coat.** Soft nub-textured wool, fully lined. Medium blue. $21.90. [$20-25] **Cardigan Tweed Coat.** Oatmeal-beige tweed of wool and nylon, braid trim, fully lined. $19.90. [$20-25] **A-line Coat.** Soft textured wool and nylon with back shoulder yoke, fully lined. Yellow. $22.90. [$25-30] Spring/Summer 1965.

Benchwarmer Jacket. Rayon and cotton poplin treated to resist stain and rain. Drawstring hood lined in acetate. Industrial-size brass zippers and rings. Burgundy wine. $5.90. [$20-25] Spring/Summer 1966.

111

Women's Fashions
Half Sizes

Evening dresses for that formal occasion now available in shorter women's half sizes. **Jacket Dress.** Acetate and rayon crepe with delicate leaf pattern sheer nylon shoulder inset. Fully lined in acetate. Light mint green. Floor length, $25.50. [$45-55] Street length, $23.50. [$35-40] **Long Draped Shoulder Panel Dress.** Embroidered midriff has rhinestones and metallic studs. Nylon tricot over acetate taffeta. Orchid. $22.50. [$45-55] Spring/Summer 1964.

Party and semi-formal occasion dresses. **Lace Jacket Dress.** Acetate and nylon lace, lined in acetate. Scalloped jacket has rhinestone ornaments. Pink. $17.84. [$45-50] **Shirtwaist Chiffon.** Rayon chiffon, dress lined in acetate taffeta. Light blue. $17.84. [$30-35] **Panel Skirt Silk Shantung Dress.** Fully lined in acetate taffeta. Elbow length sleeves. Pin not included. Apricot. $19.84. [$40-45] Spring/Summer 1964.

Three-piece suit with a ¾ length coat lined in acetate. Cotton lace blouse is edged with crochet trim. Beige and white. $18.84. [$30-35] Spring/Summer 1964.

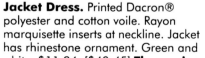

Dressy ensembles with important details such as an embroidered jacket and rhinestone ornaments. In linen-look rayon, rayon and acetate crepe, textured acetate and rayon, and cotton knit. $11.84-$16.84. [$25-30] Spring/Summer 1964.

Jacket Dress. Printed Dacron® polyester and cotton voile. Rayon marquisette inserts at neckline. Jacket has rhinestone ornament. Green and white. $11.84. [$40-45] **Three-piece Outfit.** Sleeveless overblouse and skim skirt in linen-look rayon. Jacket of linen, acetate, and cotton in open-work knit with the look of hand crocheting. Beige. $16.84. [$20-25] **Dress and Duster Set.** Printed acetate crepe dress with self-tie rope belt. Duster of rayon and acetate, lined in acetate taffeta. Green and turquoise dress, green duster. Set, $18.84. [$20-25] Spring/Summer 1964.

Dark navy blue has a slimming effect for the shorter heavier woman. The three-piece outfit on the left features a loop-trimmed jacket. Double knit Orlon® acrylic. $24.50. All other dresses from $10.84-$15.84. [$25-30] Spring/Summer 1964.

Abstract Print Jacket Dress. Crystal-pleated skirt, grosgrain belt. Jacket with ¾ sleeves and mock pockets. Arnel® jersey. Green and white. $15.84. [$30-35] **Simulated Pearl Button Coatdress.** Dainty loop closings. Light green. $14.84. [$35-40] Spring/Summer 1964.

Embroidered Cotton Eyelet Shirtdress. Simulated pearl buttons. Light blue. $14.84. [$20-25] **Tucked Bodice Dress.** Printed voile of Dacron® polyester and cotton. Light green, tan, yellow, and white on light blue. $10.84. [$30-35] **Bands of Lace Dress.** Rhinestone centered buttons. Pima cotton. Light blue. $9.84. [$20-25] Spring/Summer 1964.

A popular style is this polka-dot draped neckline dress in Arnel® jersey. The floral dress at right features a cut-out neckline trimmed with a bow. Brown with white polka dots, $10.84. [$35-40] Pink, tan, and brown floral, $9.84. [$25-30] Spring/Summer 1964.

Black and white designs in a slim jacketed style and a coachman dress. Both dresses in Arnel® jersey. Jacket dress, $14.84. [$20-25] Coachman dress, $10.84. [$15-20] Spring/Summer 1964.

These dresses sized for shorter women. **Tucked Bodice Pleated Dress.** Dacron® polyester and cotton sheer dress, rayon grosgrain belt. Honey gold on white. $10.84. [$20-25] **Princess Skimmer.** Printed rayon and flax, lined with cotton. Rose, olive, peach, and pale yellow. $10.84. [$20-25] **Tie-collar Dress.** Dacron® polyester batiste. Fine tucks on bodice front. Light rose pink. $10.84. [$18-20] **Scallop-collar Dress.** SuPima® cotton broadcloth, front bodice lined. Light copen blue. $13.54. [$25-30] Spring/Summer 1965.

Just say "Charge It" when you phone your order . . see page 637   SEARS 47

These styles were fashioned for shorter women who wear half sizes. Also available in Misses sizes. **Scroll-trimmed Dress.** Misty rayon chiffon with self-braid scroll trimming on front bodice. Fully lined with acetate taffeta. Rose pink. $19.00. [$35-40] **Lace Jacket Dress.** Acetate and rayon crepe dress with acetate and nylon lave on jacket and bodice. Skirt fully lined with acetate taffeta. Light blue. $25.00. [$40-45] Scallop Lace Evening Gown. Acetate and nylon lace bodice over a nylon sheer skirt. Fully acetate taffeta lined. Pink. $25.00. [$50-55] Spring/Summer 1965.

Glamorous lace adorns these formal occasion dresses sized for the shorter woman. **White Lace Gown.** Acetate that looks like expensive peau de soie. Bolero-effect acetate and nylon lace. Completely lined. Also available in Misses sizes. $25.00. [$25-30] **Chantilly-type Lace Jacket and Dress.** Rayon, acetate, and nylon lace, fully lined with acetate taffeta. Dusty rose. $19.84. [$40-45] Fall/Winter 1965.

115

Tattersall Checks. Dacron® polyester and rayon with knife-pleated skirt. Light copen blue and white. $13.94. [$25-30] **High-waist Dress.** "Goes on dates, goes casual." Light pink linen-look rayon bodice, light pink and yellow tattersall checked skirt and sleeve trim. $10.94. [$35-40] Spring/Summer 1966.

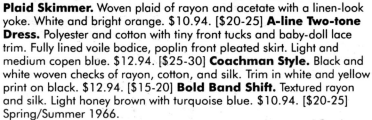

Plaid Skimmer. Woven plaid of rayon and acetate with a linen-look yoke. White and bright orange. $10.94. [$20-25] **A-line Two-tone Dress.** Polyester and cotton with tiny front tucks and baby-doll lace trim. Fully lined voile bodice, poplin front pleated skirt. Light and medium copen blue. $12.94. [$25-30] **Coachman Style.** Black and white woven checks of rayon, cotton, and silk. Trim in white and yellow print on black. $12.94. [$15-20] **Bold Band Shift.** Textured rayon and silk. Light honey brown with turquoise blue. $10.94. [$20-25] Spring/Summer 1966.

"Sissy" Shirtwaist. Dacron® polyester and cotton voile, edged with cotton lace. Light grass green with white. $14.94. [$20-25] **Misty Floral Two-piece Dress.** Pastel flowers on Dacron® polyester and cotton voile. Overblouse has pull-through tie. Fully cotton lined except for sleeves. $9.94. [$30-35] **Baby Doll Dress.** White cotton lace collar on Dacron® polyester and voile dress, fully cotton-lined. $8.94. [$25-30] **Empire Coat and Dress Set.** Dress is woven striped seersucker of acetate and cotton, coat is Dacron® polyester and cotton voile. Pink and white. $13.94. [$35-40] Spring/Summer 1966.

Dre
Dr
jers
Dr
with
Dre
$1
not

Teen Girls' Fashions

Casual Separates

Striped Turtleneck. Woven striped fabric of Dacron® polyester and stretch nylon. Red and yellow. $4.84. [$10-12] **Pullover Shift.** Wool and stretch nylon by J. P. Stevens. Red. $9.97. [$15-20] Fall/Winter 1964.

"Accessory Jackets" in abstract or flower print. In rayon/flax and rayon/silk. Each, $3.94. [$20-25] **Back-wrap or Box-pleated skirt.** Dacron® polyester and combed pima cotton. Charcoal dark brown and slate blue. $5.97. [$15-20] Spring/Summer 1964.

Country-look polka and denim fashions. **Cotton Percale Blouses.** Back zip. Blue or red. $2.97. [$20-25] **Pert Skirt.** Skirt in blue acetate and cotton denim, belt and attached shorty-pants in red with white polka dots. $5.97. [$25-30] **Culotte.** Stitched pleat. $4.97. [$20-25] Spring/Summer 1964.

Young Juniors separates in burgundy and pink. **Bermuda Collar Shirt.** Dacron® polyester and cotton oxford cloth. $3.97. [$8-10] **Knit Vest.** Wool and mohair. $4.97. [$12-15] **Sheath Skirt.** Wool flannel, fully lined in rayon. $6.97. [$8-10] **Pleated Plaid Jumper.** Wool plaid, blouson bodice lined. $14.97. [$15-20] Fall/Winter 1965.

Young Junior sweaters with a touch of mohair. **Italian Import Cable Knit Cardigan.** Wool/mohair/nylon knit. Yellow. $8.94. [$10-12] **Deep V Pullover.** Mohair/wool/nylon knit in loopy texture. Shown worn over blue sleeveless shell. Frosted bright blue pullover, $7.94. [$15-20] Sleeveless shell, $4.94. [$10-12] **Textured Cardigan.** Mohair and wool knit. White turtleneck not included. Pink. $7.94. [$10-12] **Big Collar Pullover.** Wool and mohair knit in big bold stripes. Blue and medium brown. $8.94. [$10-12] **Knit Headband.** Shown in yellow, white, and light blue. Mohair and Orlon® acrylic with brushed texture. Elastic back for better fit. $.94 each. [NPA] Fall/Winter 1964.

Teen Girls' Fashions

Sportswear and Beachwear

A collection of knit swimsuits in stretch nylon. The striped two-piece is Orlon® acrylic/cotton/rubber. $6.94-$7.94. [$35-55] Spring/Summer 1964.

Knit swimsuits for teen girls. **Two-piece Stretch Nylon.** Belt and camisole top in multicolor print on white. $7.94. **Black Nylon Tank.** $5.94. Spring/Summer 1964.

Floral designs in bright yellow. **Sportswear.** Tops, $2.94-$3.94. [$15-20] Shorts, $3.97. [$12-15] Capri Pants, $4.97. [$35-40] **Matching Beachwear.** Two-piece swimsuit in cotton duck. $5.94. [$50-55] Shift, $3.94. [$20-25] Spring/Summer 1964.

Teen Girls' Fashions

Outerwear

Ski togs for the sporty teen girl. **Hooded Quilted Tunic Jacket.** Bleached white lamb fur and water-repellent nylon taffeta quilted to Kodel® polyester fiber fill. Blue with black vinyl trim and white fur. $17.90. [$45-50] **Multicolor Print Reversible Jacket.** Concealed hood. Acetate surah print reverses to solid color royal blue nylon taffeta quilted to polyester fiber fill. Water repellent. $12.70. [$20-25] Fall/Winter 1964.

Casual Suit. Yellow textured rayon, concealed buttons on jacket front. Price includes 3-cent Federal Excise Tax on decorative pin. $9.94. [$40-45] **A-line Coat.** Yellow textured rayon laminated on polyurethane foam, acetate lined. Two concealed pockets. $19.94. [$50-55] **Accessories.** Pillbox beret and bag. Price includes a 10-cent Federal Excise Tax on bag. $4.47. [$20-25] Flower cropped Breton hat in rayon organza. Yellow. $4.47. [$20-25] Spring/Summer 1964.

The suit and patch pocket coat are Dacron® polyester and cotton seersucker that looks like linen. In black and white plaid. Suit, $11.76. [$25-30] Coat, $23.76. [$25-30] Spring/Summer 1964.

Checked Suit. Woven navy blue and white rayon and acetate, white rayon collar. Lined jacket and seat-lined skirt. $16.94. [$30-35] **Deerstalker Hat.** White cotton pique, rayon lined. Rayon grosgrain bow at top. $4.97. [$15-20] **Skimmer Coat.** Foam-laminated flannel of rayon and acetate, detachable white cotton pique collar, fully lined in acetate taffeta. Interesting off-centered buttoning. Light gray heather. $18.94. [$20-25] **Knit Coat.** Foam-laminated acrylic coat with contrasting trim, lined. Pink and white. $21.94. [$20-25] **Houndstooth-checked Coat.** Woven wool and nylon, lined in acetate taffeta. Features white leather-look vinyl buttons and low-slung belt. Navy blue and white. $24.94. [$25-30] Spring/Summer 1966.

Assorted rainwear for Juniors. **Plaid Chesterfield.** Blue and olive cotton plaid with black velvet collar, acetate taffeta lined. Zip-liner of acrylic pile. $16.94. [$15-20] **Multi-gored Poplin.** Polyester and rayon, acetate taffeta lined. Medium copen blue. $19.94. [$15-20] **Multi-pocketed Trench.** Polyester and combed cotton poplin, acetate taffeta lined. Peach. $18.94. [$15-20] **Chelsea-look Floral Print.** Orange and white flowers on black rayon, orange acetate taffeta lining. Zip-liner of acrylic pile. Striped belt of rayon and cotton faille. Head kerchief included. $24.94. [$45-50] Spring/Summer 1966.

The Poncho. Wool melton-cloth, trimmed with contrasting wool braid. Unlined. White with black trim. $19.70. [$25-30] **Suede-look Coat.** Cotton fabric imported from Spain, contrasting leather-look trim, lined with acrylic pile. Light fawn brown with dark brown trim. $26.90. [$20-25] Fall/Winter 1965.

Water-repellent wet look jackets. **Pullover.** Drawstring front nylon taffeta. Electric blue. $4.97. [$25-30] **Zip-front Floral.** Water-repellent cotton, hood lined in pink cotton poplin. Pink on black. $5.97. [$25-30] **Zip-front Polka Dot and Stripes.** Polished cotton, polka dot cotton hood lining and pockets. Black and white. $4.97. [$30-35] Spring/Summer 1965.

129

The "navy" look for little girls in cotton gabardine. The coat and dress ensemble arrived boxed. Red, white, and blue. $10.99. [$40-45] Padre hat in fine straw, $2.99. [$20-25] Nylon gloves, $1.00 pair. [$5-8] Spring/Summer 1965.

Nautical wear for bigger girls, all in cotton twill. **Sailor Duster and Dress.** $11.97. [$40-45] Coatdress. Navy braid trim in front and back. $7.97. [$40-45] **Sailor Dress.** $6.97. [$30-35] Spring/Summer 1965.

144

A collection of pink and white dresses show the popular shapes and styles for girls. **Gingham Border Jacket and Dress.** Button back jacket and sleeveless dress of white cotton pique. $5.90. [$25-30] **Chelsea Collar Jumper and Blouse.** Both pieces have button back closing. Jumper in textured cotton and rayon. Blouse of acetate crepe. $5.90. [$30-35] **A-line Dress.** Linen-look textured cotton with multicolor schiffli-embroidered border design. $4.90. [$25-30] **Bow-trimmed Dress.** White textured combed cotton with schiffli-embroidered butterfly border in front. Buttons in back. $5.90. [$30-35] **Chelsea-collar Duster and Dress.** Linen-look rayon. $5.90. [$25-30] Spring/Summer 1965.

Slim style jumper dresses for girls. **Plaid V-neck Pullover.** Brown, beige, rust, and green cotton plaid, beige cotton broadcloth blouse. $2.99. [$15-20] **A-line Pullover Jumper.** White cotton poplin, lilac and white woven check cotton gingham blouse. $2.99. [$15-20] **Jumper-effect Dress.** One-piece with a two-piece look. Red and white woven checks of Arnel® triacetate and cotton. White cotton collar and sleeves. $2.99. [$20-25] Spring/Summer 1965.

Beautiful fashions for the dress-up occasion. **Cotton Lace Jacket and Dress.** Fully lined. Dress has polished cotton cummerbund. White and blue. $8.90. [$30-35] **Schiffli-embroidered Dress.** Scalloped bodice buttons in back, pleated cummerbund in front. Cotton broadcloth. $5.90. [$30-35] **Lace Duster and Dress Ensemble.** Duster of white cotton lace, fully lined. Skimmer dress of peacock blue textured rayon and silk. $8.90. [$40-45] Spring/Summer 1965.

The jumper look for girls. **Classic Plaid.** Red, white, and blue woven plaid in Fortrel® polyester and cotton. $5.94. [$15-20] *Inset.* Shown in solid colors. **Princess Style.** Cotton twill. Red with contrasting stitching. $4.94. [$20-25] **Nautical Look.** Cotton twill. Collar with braid trim. Navy blue. $4.94. [$20-25] Spring/Summer 1965.

145

The sophisticated look in various shades of gray, white, and gold. **Striped Dress.** Looks like a jumper. Woven textured cotton in dark charcoal gray and antique gold, white rayon collar and sleeves. $8.94. [$20-25] **Knitted Dress.** Orlon® acrylic with bold-rib knit turtleneck and sleeves. Pullover style, with tie belt. Medium gray with gold color trim. $10.94. [$15-20] **A-line Dress.** Charcoal gray and white trimmed with gold. Cotton and acetate blend that looks like flannel, knit yoke and sleeves of acrylic rayon. $8.94. [$15-20] Fall/Winter 1965.

Schiffli-embroidered Two-tone Dress. Cotton knit with embroidered bodice. White with Royal blue. $5.94. [$20-25] **Scallop Embroidered Jacket Dress.** Linen-look rayon and cotton in bright rose. $6.94. [$25-30] **Rick-rack Trim Jumper and Blouse.** Brushed flannel jumper of rayon and acetate, acetate crepe blouse. $4.94. [$20-25] **Shift Jumper and Blouse.** Cotton corduroy jumper with multicolor floral printed border. White cotton broadcloth blouse. $5.97. [$20-25] Fall/Winter 1965.

Beautiful fall color combination of red, camel tan, and brown. **Schiffli-embroidered.** Cotton broadcloth dress with wheat design embroidery. $5.94. **Pleated Skirt Dress.** Polyester and cotton. $5.94. [$25-30] **Embroidered Front A-line Dress.** Rayon flannel with schiffli-embroidered front panel. $4.94. [$25-30] Fall/Winter 1965.

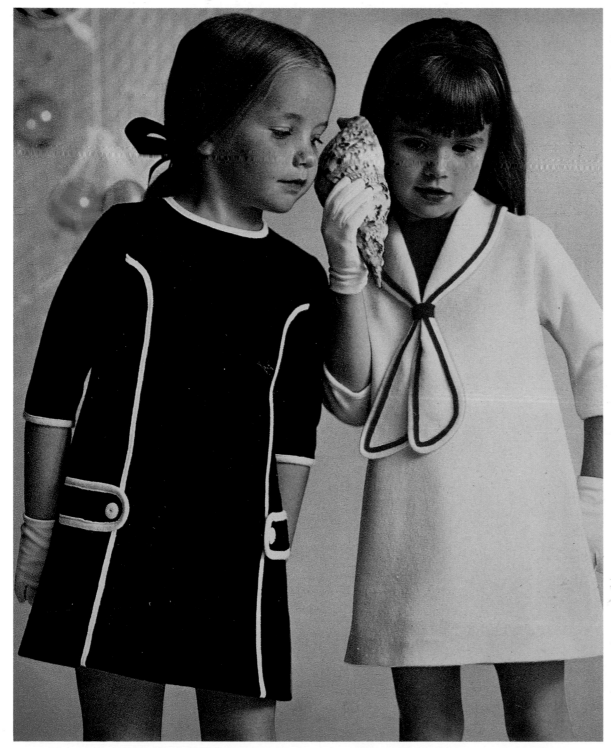

Sears best designer dresses of Orlon® acrylic double-knit are presented here in a navy couture-inspired dress with white trim, and a classic sailor-look white dress with red dickey and trim. Each, $10.99. [$20-25] Spring/Summer 1966.

Cotton Gingham Smock. Buttons in back. Pink and white check. $3.44. [$15-20] **Mod-look Dress and Scarf.** Blue bandanna printed cotton broadcloth, trimmed with rick-rack. $3.44. [$15-20] **Sailor-Tie Dress.** Cotton pique. Navy blue collar with polka-dot trim. Yellow. $3.94. [$15-20] Spring/Summer 1966.

Flare top sun suits, tennis dresses, and bubble sunsuits, all in cotton. $1.99 each. [$20-25] Spring/Summer 1965.

Recreational and casual wear in cotton seersucker. Left. **Culotte and Jamaica Shorts.** Belts included. Blue and white. $1.97-$3.94. [$15-20] **Matching Bikini Triangle.** Stripes reverses to solid white. $.94. [$10-12] Right. **Scallop Edge Two-piece Swimsuit.** Plastic belt included. Pink and white. $3.97. [$45-50] **Parka.** Pink and white. $3.97. [$20-25] **Cardigan Style Shift.** Trimmed with contrast rick-rack. Pink and white. $2.97. [$15-20] Spring/Summer 1964.

Girls-Sportswear. Popular cotton poplin shift set with matching striped panty. Turquoise and white. $1.99. [$18-25] Spring/Summer 1965.

The nautical look in red white and blue cotton poplin. The flouncy skirt shift shown lower left was called a "discotheque swing shift." $2.99-$5.99. Spring/Summer 1965.

The Skirted Look. Red and white gingham check top of stretch nylon knit. White Arnel® triacetate sharkskin skirt has attached gingham check panties. $5.94. [$20-25] **Double-breasted Look One-Piece.** Trimmed with braid and tarnish-proof anchor buttons. Front bodice lined. Red and white stretch nylon knit. $5.94. [$20-25] **Seersucker Two-piece.** Cotton seersucker with matching kerchief. Lined trunks. Pink and white stripes. $3.94. [$20-25] Spring/Summer 1965.

Little girls' styles were fashioned after teen and adult styles. The seasonal favorite, gingham checks and lace, is presented here in pink and white cotton coordinates. **Pullover Shift.** $2.97. [$20-25] **Sleeveless Overblouse.** $1.97. [$15-20] **Jamaicas.** Complete lined in cotton. $2.47. [$15-20] **Roll-sleeved Blouse.** $2.97. [$15-20] **Skirt.** $3.97. [$15-20] Spring/Summer 1965.

153

Children's Fashions

Boys' Wear - Sportswear

Below:
Boys' sweaters in two popular styles, "fashioned after the big boys." The cadet cardigan features a zip-up front with a novelty metallic pull. The blazer stripe cardigan has a ribbed knit border trim all around. Polyester, mohair, and wool blend. Each, $4.99. [$25-30] Fall/Winter 1965.

Above: Playwear for active boys. **Junior Handyman One-Piece Coveralls.** Rugged 100% cotton Fisher Herringbone, $3.47 [$35-45], or cotton with extra knee patches, $1.99. [$35-40] **New Space Commander Jump Suit.** 100% heavyweight cotton twill. Blue. $3.99. [$125-150] **Authentic-style Army Fatigue Outfit.** For "backyard commandos" made from Army surplus 100% cotton sateen. Olive drab color. Jacket has insignia and name tag with metal buttons. Cap included. Belt not included. $6.77. [$95-125] **Complete 5-piece Baseball Uniform.** Gray cotton flannel shirt and pants with dark gray stripes. Blue gabardine cotton cap, blue cotton socks. Press-on letters for team name. Belt not included. $5.87. [$125-175] An additional official Little League team cap could be purchased for just $.94 more. [$55-65] Spring/Summer 1964.

Sears collection of bleeding madras shirts were made domestically. Shown in traditional Ivy-style, henley collar pullover style, and the popular Jac-shirt. 100% cotton. $1.97. [$20-25] Spring/Summer 1966.

The Award Sweater. Bulky-knit Acrilan® is guaranteed for one year or receive a free replacement. $7.94. [$75-95] **Authentic School Letter.** Deep-tufted wool chenille. State any letter except "X". $1.00. [NPA] Fall/Winter 1964.

Sweaters in a shaggy blend of Kodel® polyester/mohair/wool. **Button Cadet.** Gold coin-look buttons on contrasting trim. Green heather and black. $6.90. [$55-65] **Zipper Cadet.** Shown in Gray and Red heather (far right). $6.90. [$55-65] **Scandinavian-look Zip Cadet.** Blue jacquard pattern on white. $6.90. [$55-65] **V-neck Pullover.** Navy. $5.90. [$20-25] **Classic 5-button Cardigan.** "So good looking it can double for a sport coat. Fine for casual wear too." Camel tan. $6.90. [$20-25] Fall/Winter 1964.

159

Children's Fashions

Boys' Wear - Outerwear

Assorted boys' jackets. **Nylon Popover.** In 6 colors. Shown in Electric blue. $4.97. [$30-35] **Cotton Casual.** Lined in batik-type cotton print. Shown in Medium blue and Light olive. $6.94. [$30-35] **Reversible Jackets.** Combed cotton plaid reverses to cotton twill solid. British style stand up collar on popular woven plaid, $5.94. [$30-35] Cadet-style collar on woven American madras plaid, $5.97. [$30-35] Spring/Summer 1964.

"Wear your school colors and letters!" Athletic "letter" jackets came in two styles. **Part-woolen, Quilt-lined, Leather-look vinyl sleeves.** Quilted acetate lining. $8.80. [$85-115] **Part-woolen Reverses to Lustrous Satin.** Reverses to solid color acetate satin. Two pockets on each side. $8.80. [$85-115] **Press-on White Letter.** Cotton felt. State any letter but "X". $.15. Fall/Winter 1964.

Flannel-lined Cotton Sheen Jackets. British style storm collar. $3.97. [$45-50] **Athletic-style Reversible Jacket.** Cotton sheen with cotton knit collar, cuffs, and waistband. Reverses to solid color. $5.94. [$55-65] Press-on white cotton felt letter is 5" high and is available in any letter but "X" for an additional 15 cents. **Baseball-style Jacket.** Each has ten emblems of the American League and National League teams. Cotton poplin body, lined in cotton flannel. Knit collar, cuffs, and waistband. Water-repellent treated. $4.94. [$75-95] Spring/Summer 1964.